T0134972

Sustainable Civil Infrastructures

Editor-in-chief

Hany Farouk Shehata, Cairo, Egypt

Advisory Board

Khalid M. ElZahaby, Giza, Egypt
Dar Hao Chen, Austin, USA

Sustainable Infrastructure impacts our well-being and day-to-day lives. The infrastructures we are building today will shape our lives tomorrow. The complex and diverse nature of the impacts due to weather extremes on transportation and civil infrastructures can be seen in our roadways, bridges, and buildings. Extreme summer temperatures, droughts, flash floods, and rising numbers of freeze-thaw cycles pose challenges for civil infrastructure and can endanger public safety. We constantly hear how civil infrastructures need constant attention, preservation, and upgrading. Such improvements and developments would obviously benefit from our desired book series that provide sustainable engineering materials and designs. The economic impact is huge and much research has been conducted worldwide. The future holds many opportunities, not only for researchers in a given country, but also for the worldwide field engineers who apply and implement these technologies. We believe that no approach can succeed if it does not unite the efforts of various engineering disciplines from all over the world under one umbrella to offer a beacon of modern solutions to the global infrastructure. Experts from the various engineering disciplines around the globe will participate in this series, including: Geotechnical, Geological, Geoscience, Petroleum, Structural, Transportation, Bridge, Infrastructure, Energy, Architectural, Chemical and Materials, and other related Engineering disciplines.

More information about this series at http://www.springer.com/series/15140

Mohamed Meguid · Erol Guler
J. P. Giroud
Editors

Advances in Geosynthetics Engineering

Proceedings of the 2nd GeoMEast
International Congress and Exhibition
on Sustainable Civil Infrastructures,
Egypt 2018 – The Official International Congress
of the Soil-Structure Interaction Group
in Egypt (SSIGE)

 Springer

Editors
Mohamed Meguid
Faculty of Engineering
McGill University
Quebec, QC, Canada

J. P. Giroud
National Academy of Engineering
Paris, France

Erol Guler
Bogazici University
Istanbul, Turkey

ISSN 2366-3405
ISSN 2366-3413 (electronic)
Sustainable Civil Infrastructures
ISBN 978-3-030-01943-3
ISBN 978-3-030-01944-0 (eBook)
https://doi.org/10.1007/978-3-030-01944-0

Library of Congress Control Number: 2018957285

This Springer imprint is published by the registered company Springer Nature Switzerland AG
The registered company address is: Gewerbestrasse 11, 6330 Cham, Switzerland

Contents

About the Editors

Dr. Meguid received his Ph.D. degree in Geotechnical Engineering from the University of Western Ontario. He spent one year as a postdoctoral fellow at Queen's University before accepting an industrial position as a project manager for a prominent consulting company in Canada. In 2004, he joined McGill University as an Assistant Professor of Civil Engineering. During his tenure at McGill, he held numerous administrative positions, including Associate Chair of the Civil Engineering Department between 2010 and 2012 and Associate Dean of the Faculty of Engineering between 2012 and 2016. He is currently the Chair of Civil Engineering Department at McGill University.

The teaching interests of Dr. Meguid include numerical methods in geomechanics, analysis and design of subsurface structures and sustainable geotechnique. His research can be broadly categorized into three primary areas of specialization (i) geosynthetic engineering, (ii) sustainable subsurface infrastructures, (iii) multi-scale modeling of granular material in contact with solid structures. He has been consulted in several national and international projects. He is a member of the Trenchless Engineering International Research Advisory Committee. He received several teaching and research awards, and his graduate students have been recognized with various honors from professional societies in North America.

Dr. Erol Guler is a full professor of geotechnical engineering at Bogazici University, Istanbul, Turkey, since 1989. He acted as the Director of Environmental Sciences Institute of Bogazici University between 1996 and 1999 and as the Chairman of the Civil Engineering Department between 2004 and 2010. He was a Visiting Fulbright Professor at the University of Maryland between 1989 and 1991. He is the leading geosynthetic scientist in Turkey, having been an IGS Member since 1989. He founded the IGS Turkish Chapter in 2001 and served as its president until 2005, and was reelected as a president again in 2011. He was the organizer for the first two national geosynthetic conferences in 2004 and 2006 and is currently the chairman of the 7th congress which will be held in 2017. He was also of the Chairman of the 2016 European Regional Conference of IGS, EuroGeo6. He has been a member of the International Standards Organization (ISO) Technical committee on geosynthetics as a representative of the Turkish Standards Institute since 2002. He is currently the Convener of the WG2 of ISO/TC221 (Technical Committee on geosynthetics) and is also the Convener of the WG2 of CEN-TC189 (European Committee for Standardization's Technical Committee on geosynthetics). He is currently an international member of the USA TRB Committee on Geosynthetics. His research has focused mainly on geosynthetic-reinforced walls, and specifically, he conducted research on the use of marginal soils in such structures and their behavior under earthquake loading conditions. His research work includes numerical studies as well as shaking table tests and full-scale tests. He has over one hundred scientific publications. In addition to his research work, he has extensive practical experience, including design work for numerous projects where geosynthetics were used as reinforcement or liners.

Prof. Dr. Eng. J. P.Giroud
Member of the US National Academy of
Engineering

Dr. Giroud is a consulting engineer, member of the US National Academy of Engineering, Doctor Honoris Causa of the Technical University of Bucharest, Past President of the International Geosynthetics Society (IGS), Chairman Emeritus and founder of Geosyntec Consultants, and Chairman of the Editorial Board of Geosynthetics International, Chevalier in the Order of the Légion d'Honneur and a former professor of geotechnical engineering. He has authored over 400 publications. He coined the terms "geotextile" and "geomembrane" in 1977. He has developed many of the design methods used in geosynthetics engineering and has originated a number of geosynthetic applications. In 1994, the IGS named its highest award "The Giroud Lecture," "in recognition of the invaluable contributions of Dr. J. P. Giroud to the technical advancement of the geosynthetics discipline"; a Giroud Lecture is presented at the opening of each International Conference on Geosynthetics. In 2002, he became Honorary Member of the IGS with the citation "Dr. Giroud is truly the father of the International Geosynthetics Society and the geosynthetics discipline." In 2005, he has been awarded the status of "hero" of the Geo-Institute of the American Society of Civil Engineers (ASCE) and has delivered the prestigious Vienna Terzaghi Lecture. In 2005–2006, he presented the Mercer Lectures, a prestigious lecture series endorsed jointly by the IGS and the International Society for Soil Mechanics and Geotechnical Engineering (ISSMGE). In 2008, he delivered the prestigious Terzaghi Lecture of the ASCE. In 2016, he delivered the prestigious Victor de Mello Lecture of the ISSMGE and, in 2017, the prestigious Széchy Lecture, in Budapest. He has 56 years of experience in geotechnical engineering, including 48 years on geosynthetics.

Experimental and Numerical Modelling of a Reinforced Structure

R. E. Lukpanov and Talal Awwad[✉]

Eurasian National University of L.N. Gumilyov, Astana, Kazakhstan
dr.awwad@ymail.com

Abstract. The paper presents a complex of research of the soil retaining structure. The results of a numerical analysis, as well as laboratory modelling in the 1/40 scale, are presented in this paper. The paper includes constructive solution analysis for the retaining structure construction. Solution implies the usage of piles as retaining element and geogrid as an element of soil reinforcement. A methodology for optimizing a constructive solution by using modern finite element method is also proposed. A quantitative estimate of the geogrid application efficiency for different task design within the single constructive solution of retaining structure (RS hereinafter) is given. The model tests include model testing of an RS without geosynthetic reinforcement elements (geogrid); and series of model tests of an RS with a geogrid for various piles spans as well. The test results are summarized in a graphical and tabular form. In general, the results of the research showed the effective use of geosynthetic reinforced material, as an element of soil strengthening.

1 Introduction

During the last years, many experimental, numerical, and analytical studies have been performed to investigate reinforced soil foundations – RSFs and a stability of embankment dams (Chen and Abu-Farsakh 2015; Awwad et al. 2017); to investigate the behavior of reinforced soils and foundations for different reinforcement types: by use of soil cement columns and prefabricated vertical drains (Ishikura et al. 2016; Ye et al. 2015); stone columns (Killeen and McCabe 2014; Ng and Tan 2014); Lime pile (Abiodun and Nalbantoglu 2015; Awwad 2016); lime–cement columns (Larsson et al. 2009). In these terms, there are perspective technologies associated with improvement of soils adjacent to the existing urban buildings and structures (Awwad and Donia 2016; Zhussupbekov and Zhunisov 2013; Awwad and Al-Asali 2014).

The application of geosynthetic materials for the stability of the soil embankment has a great demand and consumption in the world practice: by use of multiple-geocell or planar geotextile reinforcing layers (Tafreshi et al. 2016); by using the Geosynthetics-encased columns (Alkayyal and Awwad 2015); and by use of geogrid sheets (Chakraborty and Kumar 2014; Alkayyal et al. 2014). Today intensive usage of these materials in road construction is observing in Kazakhstan. Nevertheless, the practical implementation of geosynthetic reinforcement elements in production,

© Springer Nature Switzerland AG 2019
M. Meguid et al. (Eds.): GeoMEast 2018, SUCI, pp. 1–11, 2019.
https://doi.org/10.1007/978-3-030-01944-0_1

the issues of its effective usage in Kazakhstan is still under the question (Lukpanov 2017). To study the issue, both numerical modelling and laboratory model tests had been made.

The complex of research included the following steps:

1. Numerical modelling of reinforced and unreinforced soil embankment with the usage of a geogrid and a different stepped pile.
2. Model tests of reinforced and unreinforced soil embankment with the usage of a geogrid and a different stepped pile.

The numerical and model testing is shown in Fig. 1.

Numerical modelling   **Model test**

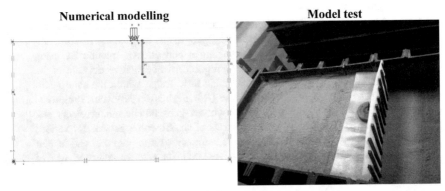

A) Model of an unreinforced soil embankment

B) Model of a reinforced soil embankment

Fig. 1. Concept of research

Precast concrete driven piles (14 m of length, 40 × 40 cm of the cross-section) were used as supporting slope structure.

Numerical simulation was carried out in Plaxis 2D, in a plane strain. When carrying out model tests, the law of dynamic similarity was used to select the model components (soil, geosynthetic, geogrid, piles, etc.) (Lukpanov 2016). Model tests were carried out at a scale of 1:40.

For the initial constructive solution, the following decision has accepted:

1. The use of piles with a length of 14 m, section 40 × 40 cm, with pile span of 1 m;
2. Application of the geogrid 150 × 150 mm, with an axial strength EA (deformation of 2%) 200000 MPa, with different pile spans;
3. Preliminary location of the geogrid is taken with a step of 1.5 m along the height of the RS.

2 Numerical Modeling

The research works by numerical modelling includes following

1. Determination of the optimal location (zone of the placement) of geosynthetic reinforced elements along the height of the RS;
2. Analysis of the joint works of geosynthetic elements at specified locations, with the aim of determining the optimal solution
3. Analysis of the application of several geosynthetic grids (from 5 to 1 with a span of 25 cm) in the chosen optimal location;
4. Investigation of the dependence of the reinforcement length adopted to reduce the number of piles (span increments).

The last task in the plain strain could be realized by the reduction of pile stiffness and increase the soil pressure.

Figure 2 (on the left) shows the locations of geosynthetic grids, with a step of 1.5 m along the height of the RS. Calculations of the slope stability of the RS have been carried out for each location separately, in order to identify the shortest length of reinforcement. Calculations showed that the optimal geolocation solution corresponds to the positions B and E, where the reinforcement length is 8 m in both.

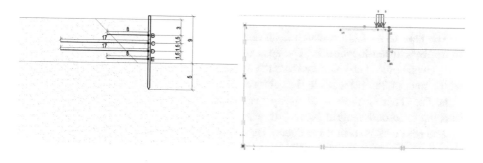

Fig. 2. Location and design scheme

Next, the joint work of the geogrid in two locations B and E had been made. The general principle of the calculation is determining the lowest total length of the geogrids (joint work B and E) under the same criteria of RS reliability and stability. The solution of the task for determination of the optimal location is represented by the following algorithm: E (n) + B (8 − n) … + B 8 − n + 1) or B (8 − n − 1)… until the stability condition of the RS is satisfied.

The results of the research showed that the optimal solution is the individual work of the geogrid either in the E location or in the B location. Since all the combinations have shown the need for a longer geogrid length (total length over 8 m). The results are presented in Fig. 3, where vertical axis - the geogrid length along E, horizon axis — the geogrid length along the B.

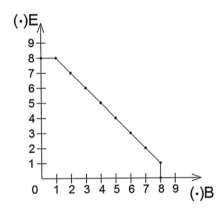

Fig. 3. Analysis of the joint work of geosynthetic reinforcement elements in locations B and E

Since the locations B and E are symmetrical, it is possible to notice identical work of the reinforced geo-elements. The point B is 3 m from the surface and the point E is 6 m from the surface. Rational solution of using the geogrid is location B. That is, from the point of view of the production technology for geogrid construction this constructive solution will be most economical.

The next task of the research is an analysis of the group work of several geogrids are nearby chosen location B. The span of the group geogrids is 25 cm (Fig. 4).

Analysis of this task showed that the application of only 10 geogrids of 5 m length (with a span of 25 cm) leads to the achievement of the necessary stability and reliability of the RS. That is, the use of group geosynthetic elements is ineffective. Therefore, choosing only one geogrid layer with a length of 8 m, located in point B (Fig. 5).

The next task is to find the dependence between the reinforcement length and piles number on the point B. This task can be realized in Plaxis by reducing the stiffness of the pile EA and increasing the soil pressure on the retaining wall.

Analysis of this task showed that the application of 10 geogrids of 5 m length (with a span of 25 cm) leads to stability of the RS. That is, the use of group geosynthetic elements is ineffective. Therefore, the last solution is using only one geogrid layer with a length of 8 m, located in point B.

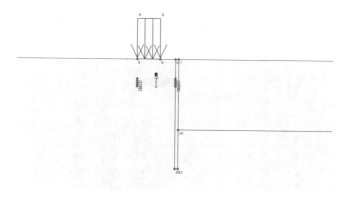

Fig. 4. Design scheme of task 3

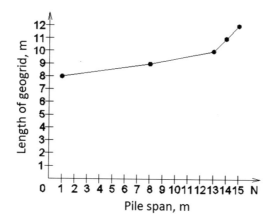

Fig. 5. Dependence geogrid length of pile span

Table 1 presents the calculated stiffness of piles and soil pressure depending on changing the pile span.

The results of the calculations showed that the increment in the length of the geogrid with increasing pile span is not significant, and therefore the most economical, in terms of material and labour, the use of the maximum pile span. However, this increase of pile span (15 m) requires additional research: an assessment of the RS facing, it`s sufficient strength between the piles and other.

3 Laboratory Modeling

The model tests included following

1. Model testing of an RS without geosynthetic reinforcement elements (geogrid);
2. A series of model tests of an RS with a geogrid for various piles spans.

Table 1. The stiffness of piles and soil pressure depending on changing the piles

No.	The span of the pile, m	EA, kN/m²	ρ, kN/m³
1	2	1500E+07	32
2	3	1000E+07	48
3	4	7500E+06	64
4	5	6000E+06	80
5	6	5000E+06	96
6	7	4200E+06	108
7	8	3750E+06	124
8	9	3333E+06	140
9	10	3000E+06	156
10	11	2727E+06	172
11	12	2250E+06	188
12	13	2307E+06	204
13	14	2142E+06	220
14	15	2000E+06	236

Each test was carried out using a static load applied to the roadbed as a point load until the ground embankment collapsed completely. The mixture was used as the equivalent material. An elastic polymeric material was used as a model of the roadway. A polymer grid with an aperture of 1.25 cm was used as a reinforced material. Flexible geotextile was used as facing element of the RS. Models of piles are made of wood covered with bitumen. A series of model tests with multilevel pile span had been made, Table 2.

Table 2. Model test of RS

No.	Test series	The span of piles, in nature (m)/in the model (cm)	
1	RS without reinforcement	2 m	5 cm
2	Reinforced RS	2 m	5 cm
3	Reinforced RS	4 m	10 cm
4	Reinforced RS	6.8 m	17 cm
5	Reinforced RS	10.6 m	26.5 cm
6	Reinforced RS	21.2 m	53 cm

The materials for the model are selected on the basis of the general law of dynamic similarity, taking into account the simultaneous action of gravity and internal stresses (Tanaka et al. 2015).

After substituting the corresponding values for the model and natural soil, we obtain the linear scale of the modelling:

$$m_c = \frac{C_M}{C_N} \times \frac{\gamma_N}{\gamma_N} = \frac{0.9}{38} \times \frac{19}{17.7} = \frac{1}{40}, \tag{1}$$

Consequently, the linear scale of the model and the object is determined by the ratio of the strength properties (cohesion) of loam and equivalent material and is equal to 1:40.

The material of the soil basement is represented by a mixture consisting of 97% fine quartz sand and 3% spindle oil by weight. The oil allows to model cohesive soils. Parameters of soils and equivalent material are presented in Table 3.

Table 3. Soil and equivalent material parameters.

The name of soils and model material	Unit weight, γ, ($\kappa N/m^3$)	Cohesion, c (κPa)	The angle of internal friction ϕ (degree)	Modulus of deformation, E (MPa)	Poisson's ratio v
Physical and mechanical parameters of the natural dam					
Loam	19.0	38	38	27	0.35
Physical and mechanical parameters of the natural dam					
Equivalent material	17.7	0.90	21	0.26	0.25

It is necessary only one parameter of reinforced element modelling is an axial strength, which can be fined by the following equation:

$$EA = T \times t \times t_g \frac{W_g}{s}, \tag{2}$$

Where EA – axial strength, kN/m; t – the thickness of reinforced element, m; T – tensile force, kN/m; tg – the thickness of geogrid rod, mm; Wg – width of geogrid rod, mm; s – space between the rods of geogrid, mm.

The axial strength of one rod of natural geogrid is 51 kN/m, then the axial strength of 1 m reinforcement is followed: 51 · 5 = 255 kN/m (where 5 is a number of rods per 1 m of the natural dam). Final parameters of the model geogrid are as followed: diameter of the rod is 0.8 mm, the cross section is $0.524 \cdot 10^{-6}$ m, and axial strength of equivalent 1/40 m of the model is EA = 100.48 kN/m, where 1/40 is a scale of natural to model dam (Lukpanov 2016).

The tests were carried out in accordance with the test program. Each test was carried out until the overall stability of the retaining structure was exhausted. Thus, a quantitative comparison of the loads (before collapse) will give a general conception of the soil embankment strengthening (percentage, compared to the unreinforced).

Figure 6 shows a picture of the collapse of the retaining structure for various tasks (without reinforcement and reinforcement with different pile spans).

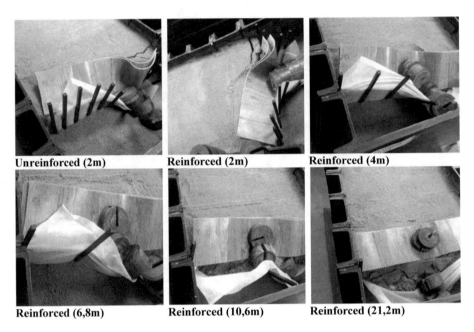

Unreinforced (2m)   Reinforced (2m)   Reinforced (4m)

Reinforced (6,8m)   Reinforced (10,6m)   Reinforced (21,2m)

Fig. 6. Modelling of the RS

The test results are summarized in a graphical and tabular form. The graph shows the dependence of the pile span and the maximum load at which the retaining structure collapsed. The table shows the comparison of absolute values and the percentage with the results of the first test (without reinforcement) (Table 4).

Table 4. Summary table of model test

Test series	Pile span, m	Absolute value, kg	Relative, % (relatively of RS without reinforcement)
RS without reinforcement	2	6.8	100
Reinforced RS	2	15.8	232
Reinforced RS	4	10.2	150
Reinforced RS	6.8	9.4	138
Reinforced RS	10.6	8.4	124
Reinforced RS	21.2	6.4	94

4   Conclusions

In general, the results of the research showed the effective use of geosynthetic reinforced material, as an element of soil strengthening (Fig. 7).

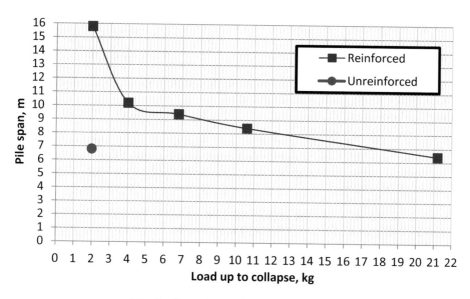

Fig. 7. Dependence of load and pile span

The results of the numerical analysis showed that the most optimal design solution would be the use of a single layer of a geosynthetic reinforcement element: 8 m long, 3 m from the surface. The results of model tests also showed the effectiveness of the geogrid. The reinforced retaining structure maintains 2.32 times the static load than the non-reinforced one. The application of the geogrid allows increasing the pile span by 10 times (21, 2/2 = 10.6). In general, the analysis of the choosing optimal pile span (with a view to reducing the material expenditure) showed a tendency to minimize costs with increasing pile span, but requires additional research, as long as a number of questions related to the choice of the facing material and its works are arise.

References

Abiodun, A.A., Nalbantoglu, Z.: Lime pile techniques for the improvement of clay soils. Can. Geotech. J. **52**(6), 760–768 (2015)

Alkayyal, H., Althoff S., et al.: Supporting structures reinforced by geogrids—the engineering challenge. In: 10th International Conference on Geosynthetics ICG 2014 "Soil Reinforcement —In-Situ Testing and Conclusions", Berlin, September 2014

Awwad, T.: Problems of Syrian historical monuments, destroyed by military action: diagnostics, strengthening and reconstruction, challenges and innovations in geotechnics. In: Proceedings of the Eighth Asian Young Geotechnical Engineering Conference, pp. 21–27 (2016)

Awwad, T., Al-Asali, B.E.: Efficiency of improving the specifications of soil lenses that are formed near the tunnel during the stages of its investment. In: Proceedings of the TC207 ISSMGE Conference, Soil-Structure Interaction, Underground Structures and Retaining Walls. Advances in Soil Mechanics and Geotechnical Engineering, vol. 4, pp. 111–118 (2014)

Awwad, T., Donia, M.: The efficiency of using a seismic base isolation system for a 2D concrete frame founded upon improved soft soil with rigid inclusions. Earthq. Eng. Eng. 15(1), 49–60 (2016)

Chakraborty, M., Kumar, J.: Bearing capacity of circular foundations reinforced with geogrid sheets. Soils Found. 54(4), 820–832 (2014)

Chen, Q., Abu-Farsakh, M.: Ultimate bearing capacity analysis of strip footings on reinforced soil foundation. Soils Found. 55(1), 74–85 (2015)

Alkayyal, H., Awwad, T.: Improvement of the dynamic behavior of soil structures underlain by liquefiable soil using the geosynthetics-encased columns. In: Proceedings of the 6th International Symposium on Deformation Characteristics of Geomaterials, Buenos Aires, Argentina, November 2015. Advances in Soil Mechanics and Geotechnical Engineering, vol. 6, pp. 792–799. IOS Press (2015)

Ishikura, R., Yasufuku, N., Brown, M.: An estimation method for predicting final consolidation settlement of ground improved by floating soil cement columns. Soils Found. 56(2), 213–227 (2016)

Killeen, M.M., McCabe, B.A.: Settlement performance of pad footings on soft clay supported by stone columns: a numerical study. Soils Found. 54(4), 760–776 (2014)

Larsson, S., Rothhämel, M., Jacks, G.: A laboratory study on strength loss in kaolin surrounding lime–cement columns. Appl. Clay Sci. 44(1–2), 116–126 (2009)

Lukpanov, R.E.: The concept of carrying out model tests of a soil embankment subject to uneven horizontal and vertical deformations. J. Vestn. ENU 4(119 (part 2)), 118–123 (2017)

Lukpanov, R.E.: Laboratory modelling of soil testing embankment reinforced by geosynthetic elements. In: 6th Asian Regional Conference on Geosynthetics, GA-2016, New Delhi, India, p. 77 (2016). ISSN 978-8-17-336390-0

Tafreshi, S.N.M., Sharifi, P., Dawson, A.R.: Performance of circular footings on sand by use of multiple-geocell or planar geotextile reinforcing layers. Soils Found. 56(6), 984–997 (2016)

Ng, K.S., Tan, S.A.: Design and analyses of floating stone columns. Soils Found. 54(4), 478–487 (2014)

Awwad, T., Donia, M., Awwad, L.: Effect of a stiff thin foundation soil layer's depth on dynamic response of an embankment dam. Proc. Eng. 189(2017), 525–532 (2017)

Tanaka, T., Zhussupbekov, A., Aldungarova, A., Lukpanov, R.: Model test on the stability of the dam model with horizontal and radial deformation of the subgrade. In: 6th International Geotechnical Symposium on Disaster Mitigation in Special Geoenvironmental Conditions, pp. 375–379 (2015)

Ye, G., Zhang, Q., Zhang, Z., Chang, H.: Centrifugal modeling of a composite foundation combined with soil-cement columns and prefabricated vertical drains. Soils Found. 55(5), 1259–1269 (2015)

Zhussupbekov, A., Zhunisov, T., et al.: Geotechnical and structural investigations of historical monuments of Kazakhstan. In: Proceedings of Second International Symposium on Geotechnical Engineering for the Preservation of Monuments and Historic Sites, Naples, Italy, pp. 779–784 (2013)

Employment of Sisal Natural Fibers as Soil Reinforcement

Hellen Fonseca[1(✉)], Rideci Farias[2], Ivonne Gongora[3],
Dra. Leidiane Garcia[4], and Matheus Viana[1]

[1] Civil Engineering, Universidade Católica de Brasília, Brasília, DF, Brazil
hevenyl@gmail.com, matheusvianadesouza@hotmail.com
[2] UCB/Reforsolo Engenharia/UniCEUB/IesPlan, Brasília, Brazil
rideci.reforsolo@gmail.com
[3] UCB/UNB/UniCEUB, Brasília, Brazil
ivonnegg_86@yahoo.com
[4] Ambiental Engineering, UCB, Brasília, DF, Brazil
leidiane.engenharia@gmail.com

Abstract. The use of natural fibers as a reinforcement for soils is a technique known for a long time by humanity. This practice has a great performance as a reinforcement material for has a high tensile strength, promoting improvement in the mechanical properties of the soils, conferring gain of mechanical resistance and reduction of his compressibility. The performance of the mechanical behavior of clay-silt soil at the Brasília Landfill, Brazil is analyzed by reinforcing the soil with natural sisal fibers with lengths of 75, 50, and 25 mm for a 0.5% fiber content of the total volume of dry soil. The aim to finding the optimal fiber length for the analyzed percentage. The mechanical strength conferred to the soil when applying such reinforcement is analyzed using the California Bearing Ratio and simple compression tests, and the results are compared with those obtained in natural soil. The plastic deformation and the reduction of the voids index are analyzed through the consolidation tests for each sample. The 75 mm sisal fiber was expected to be more viable for the reinforced soil, but it presents great difficulty of homogenization with the soil and high agglomeration of the fibers, making it difficult to evaluate. However, the 50 mm fiber presented higher mechanical strength and greater readiness for homogenization with the soil for the 0.5% content analyzed.

Keywords: Fiber reinforcement materials · Sisal fiber · Silty-sand soil

1 Introduction

In civil engineering, soil is considered a multiphase medium composed of solid particles, air, and water. The physical-chemical interaction between these phases results in a complex material, which requires different analyses for an understanding of its behavior. In certain types of projects, the soil strength at project sites may not meet the design strength required to support applied loads, for being low resistance soils, that is, soils with less load carrying capacity and more susceptible to deformation. In paving projects, one way to deal with the problem is to remove the layer of low resistance soil

© Springer Nature Switzerland AG 2019
M. Meguid et al. (Eds.): GeoMEast 2018, SUCI, pp. 12–24, 2019.
https://doi.org/10.1007/978-3-030-01944-0_2

and replace it with another suitable material. However, this practice can cause environmental problems associated with the improper disposal of materials as well as the high costs of excavating and transporting of the removed material.

For a portion of low-resistance soil, another solution is the improvement of the mechanical properties of the soil through the use of natural fibers. In this way, several studies on alternative fibers, particularly vegetable fibers, have been conducted to obtain improved the behavior of these soils in an economical and sustainable manner. Experimental studies confirmed that the addition of fibers modifies soil behavior, resulting in a more ductile, cohesive, and slightly more compressible material (Bueno 1996). Gray and Ohashi (1983) studied the mechanics of fiber reinforcement in cohesion soils and showed that the inclusion of fibers increase peak shear strength and ductility of soils under static loads.

This research will verify through laboratory tests realized with lateritic soil extracted from Brasília Landfill (Brazil), in the region of Samambaia, the improvement of soil in the presence of sisal fibers with sizes of 25, 50, and 75 mm at 0.5% fiber content in relation to the total volume of dry soil. It will also be done the comparison of the obtained results of the fibers with the samples of natural soil to verify the resistance gain, the feasibility of adding fibers to the soil as a reinforcement element and their application to sub-base and sub-grade layers in paving projects. The established fibers lengths and defined fiber content were obtained according to Trindade (2016).

Therefore, through laboratory analysis, the behavior of the soil in the presence of sisal fibers, the comparison of the properties obtained from soil tests with and without the addition of fibers, what is the optimal fiber sizes of 25, 50, and 75 mm at 0.5% fiber content relation to the total volume of dry soil, were determined. In addition, the determination of resistance characteristics of the mixtures elaborated, were obtained through California Bearing Ratio (CBR) and simple compression tests.

2 Methodology

The physical and mechanical characteristics of the soil and fibers were obtained through laboratory testing at Universidade Católica de Brasília, in the region of Águas Claras, Brazil. The soil samples were collected from the Brasilia Landfill located in the region of Samambaia Sul DF 180 - Km 51, with coordinates 15°52′05″S and 48°09′40″W. The process of soil collection was conducted in accordance with the specified Brazilian regulatory standard, *Associação Brasileira de Normas Técnicas* (ABNT)-NBR 9604/1986. The visual tactile test required for a primary assessment was conducted for a better identification of the soil type, as well as to better establish the preparation of the characterization tests, the results of which revealed the characteristics of the sandy soil. This procedure was performed in accordance with ABNT-NBR 7250/1982.

The sisal fibers were extracted from ropes made from sisal acquired in Brasilia, Brazil. While unwinding the ropes, the residues between the fibers were removed, and the fibers were cut into lengths of 25, 50, and 75 mm. Figure 1(a, b, and c) shows the fiber lengths used.

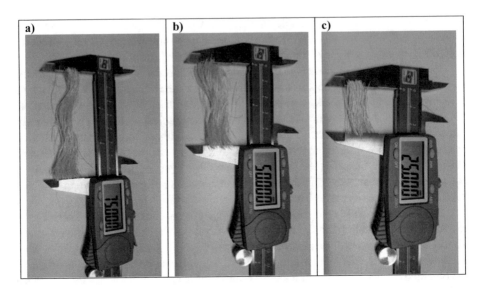

Fig. 1. Sisal lengths (a) 75 mm; (b) 50 mm; (c) 25 mm

Tests on the soil, sisal fibers, and soil-fiber mixtures were conducted as follows.

2.1 Grain Size Analysis

According to Sampaio (2010), grain size analysis is conducted to evaluate grain size distribution and determine the equivalent diameters of solid particles in soil along with the proportion of each constituent fraction of soil in relation to the weight of the soil in a dry state. Grain size analysis was conducted through sedimentation and sieving. The methodology employed for this test and the equipment used, were based on ABNT-NBR 7181/1984.

2.2 Atterberg Limits

The determination of consistency limit, also known as Atterberg Limits, is given empirically from laboratory test results, have been standardized by Arthur Casagrande.

The procedures and materials used to perform the consistency of soil limit tests were prepared according to regulatory standard ABNT-NBR 6459/1984, for determining the liquidity limit, and ABNT-NBR 7180/1984 to determine the plasticity limit.

According to Murthy (2009), plasticity index classifies the soil according to plasticity, that is, the moisture content necessary for the soil to pass from plastic state to the liquid state. The plasticity index (PI), that is use to classify the soil, is obtain by subtracting the liquid limit (LL) by the plasticity limit (PL).

The preparation of the soil to perform these tests proceeded as explained in ABNT-NBR 6457.

2.3   Specific Gravity of Soil

The specific gravity of soil was obtained by using a pycnometer. The methodology and equipment used in this test has made according to the established by the ABNT-NBR 6508/1984.

2.4   Sisal Specific Gravity

The Pentapyc 5200e – Automatic Density Analyzer, determined sisal specific gravity. The test was made at Universidade de Brasília geotechnical laboratory. The performance of the test, as well as the equipment used, were conducted in accordance with the regulatory standard D5550-00 (ASTM, 200). The apparatus has been programmed by perform the reading 10 times under an accuracy of 0.0050% and standard deviation of 0.370%.

2.5   Proctor Compaction Test

The sensitivity to compaction was evaluated through a Proctor test developed in the field of civil engineering. This test determines the adequate moisture required to achieve maximum soil compaction in road construction projects (Vargas 1977). This can be understood based as the specific humidity level at which the soil is more susceptible to compaction.

For the soil in question, the procedures and execution of the compaction test were applied as indicated by ABNT - NBR 7182/1984.

A small metal compaction mold was used owing to the integral passage of the soil sample through a 4.8 mm sieve, as specified by the standard. The compaction was conducted using a large hammer, as specified in ABNT-NBR 7182/1984, at intermediate Proctor energy of 26 blows for every three layers of the material. This compaction energy was chosen when considering the intermediate of the pavement layers.

During the compaction, it was more difficult to perform the procedure with the longer fibers.

2.6   California Bearing Ratio Test

The CBR test represents the soil support capacity as a function of the resistance to the penetration of a rod 5 cm in diameter into a layer of crushed stone, which is considered as the standard CBR of 100% (Machado et al. 2006). It also provides the soil expansion beneath a pavement when saturated, as well as indications of soil loss from the resistance caused by this saturation, as indicated in the standard ABNT-NBR 9895/1987.

The methodology applied to execute the CBR test, as well as the equipment used, are based on ABNT-NBR 9895/1987.

The resistance of the test indirectly combines the cohesion with the friction angle of the material.

2.7   Simple Compression

According to the executive branch of the Brazilian states, the *Departamento Nacional de Estrada e Rodagem* – DNER (004/1994), the compressive strength of a cohesive soil was obtained by the pressure value corresponding to the load that breaks a cylindrical specimen of soil submitted to axial loading. The relation between the decrease of height of the specimen by the application of the load and its initial height also measures the specific deformation.

The test was performed by the piston penetration control. The axial and horizontal deformations were measured with extensometers adapted to the simple compression machine.

The DNER-IE 004/94 presents the calculation of the shear strength or cohesion of the soil tested being half of the compressive strength obtained, as showed in below

$$c = \frac{R}{2}.$$ (1)

2.8   One-Dimensional Consolidation Test

The data from the consolidation test are used to estimate the magnitude and rate of both the differential and total settlement of a structure or earthfill, based on ASTM D 2435 (1996).

According to Murthy (2009), such a consolidation test is characterized by the speed and magnitude of the deformations applied to the specimen when it is laterally confined and axially loaded and drained. This test represents the conditions under which the loads applied to the surface cause soil deformation solely through compression, without any lateral deformation. Based on the initial height of the test piece, the settlements can be calculated as a function of the vertical stresses.

The ring used for testing was fixed to the base, and consolidation was performed without flooding. The coefficient of densification was determined according to the Casagrande method. The pre-consolidation pressure was established according to the method developed by Silva. The increments and decrements used for the test were 50, 100, 200, 400, and 800 kPa. The increments were made every 24 h while the decrements were applied in an analogous manner.

The test procedures, as well as the equipment and calculations used were made according to ABNT-NBR 12007/MB-3336.

3   Results and Discussion

3.1   Grain Size Analysis Test

Figure 2 shows the curve of the grain size analysis of the studied soil. Obtaining the grain size curve of the soil is importance when analyzing grain distribution and identifying the diameter for classification (Pinto 2006).

Using a graph of the grain size analysis it is possible to see the well-graded distribution of the grains.

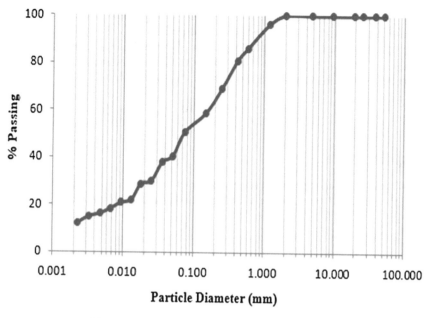

Fig. 2. Curve of grain size analysis of studied soil

According to Table 1, it could to note that the evaluated soil can be classified as coarse soil. However, based visual tactile analysis, the results of Atterberg limits, as well as a slight difference of only 0.7% between the value retained and the passing of the soil, the soil is classified as a fine material with particles of fine sand classified as an SM (silty sand), according to the Unified Soil Classification System.

Table 1. Sieve analysis of soil

Opening of sieves (mm)	% Retained	% Passing
1.19	96.2	3.8
0.59	86.1	13.9
0.42	80.9	19.1
0.25	68.9	31.1
0.149	58.7	41.3
0.075	50.7	49.3

3.2 Atterberg Limits

The average of the results found in the Liquid Limit test corresponds to value of LL = 44.47%.

The Plastic Limit test presented mean values of 34.99%.

The Plasticity Index obtained according to Eq. 2, is 10.07%, therefore, the soil has medium plasticity, according to the classification of Jekins.

Through the system of classification proposed by American Association of State Highway and Transportation Officials (AASHTO), the soil can be classify as belonging to group A7, with subgroup A-7-5, because LL-30 = 14.47 > PI

$$PI \leq LL - 30. \tag{2}$$

3.3 Specific Gravity of Soil

The result obtained after grain specific mass test with the calibrated pycnometer, was 2.5 g/cm^3, whose value is within the classification range for common soils, according to ABNT.

3.4 Sisal Specific Gravity

The results report generated by penta pycnometer are present in Table 2.

Table 2. Results of the sisal specific gravity

Data	Result	Unity
Cell size	Average	-
Cell volume	59.8259	cm^3
Temperature analyzed	25.3	C
Average volume	2.4566	cm^3
Average density	1.8649	g/cm^3
Coefficient of variation	0.4079	%

3.5 Proctor Compaction Test

Analyzing the compaction graph with the sisal and natural soil mixtures it can be seen that, the larger the fiber size is, the lower its optimal moisture content, which is due to the absorption of water because it is non-inert material, and the higher specific maximum dry mass of the analyzed content, because fiber expansion increases the total volume of the sample. Figure 3 shows compaction graphs of the mixtures and natural soil.

3.6 California Bearing Ratio Test

Figure 4 and Table 3 shows the results of the CBR obtained from samples of natural soil and soil with sisal fibers with lengths of 75, 50, and 25 mm for an evaluated content of 0.5% fiber in terms of the total volume of soil. Based on the results obtained from the California Bearing Ratio test, it was observed that the soil-fiber mixture with the highest penetration resistance was 50 mm. The 75 mm fiber was expected to

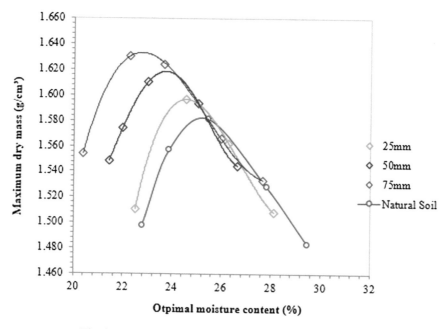

Fig. 3. Compaction graphs of the mixtures and natural soil

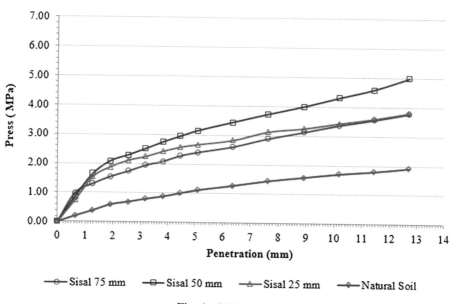

Fig. 4. CBR curves

Table 3. CBR values and expansions

Samples	CBR (%)	Expansions (%)
Natural soil	10.5	0.01
25 mm sisal fiber	28.29	0.18
50 mm sisal fiber	31.83	0.26
75 mm sisal fiber	24.23	0.40

exhibit greater strength during this test, but because of the difficulty of homogenizing the fiber with the soil, the penetration resistance could not be adequately assessed through conventional compaction procedures.

According to the results presented in Fig. 4, for 0.5% sisal fiber content, 50 mm is the optimal fiber length that provides the soil with a resistance increase of 203.14% compared with natural soil.

Table 3 shows the results of the mean expansion indices of each sample tested. According to Donisete (2016), the indexes of expansion directly affect the design of the floor and pavement, and such an evaluation is necessary because potentially expansive soils can cause irreparable pathological manifestations.

According to the executing agency of the Brazilian Ministry of Transport, the *Departameto Nacional de Infraestrutura de Transporte* (DNIT), with regard to the soil expansions found, all soil-fiber and natural soil mixtures can be used for paving layers because the lowest DNIT values allowed is that of a reinforced sub-base, whose expansion is $\leq 1\%$.

3.7   Simple Compression

Figure 5 shows the graph of the results of the peaks obtained from a simple compression and its respective cohesions. The presence of the fibers increased the initial resistance and increased the post-peak deformation. The results of the simple compression test are similar to those of the CBR terms of resistance to the soil, with the 50 mm fiber lengths being the largest among the soil-fiber and natural soil mixtures considered. The same holds for the cohesion.

The increase in the lengths of fiber in the soil increases the resistance and cohesion compared with natural soil. The 50 mm fiber length demonstrates a 150% increase in compressive resistance and fiber cohesion compared with natural soil.

3.8   One-Dimensional Consolidation Test

Table 4 shows the preconsolidation pressure of each sample. It is shown that as the length of the sisal fiber increases, the preconsolidation pressure decreases.

Table 5 shows the initial void index of each sample. The presence of fibers reduced the void indices compared with natural soil, the longer the fiber length, the greater the reduction of voids. However, a difference of less than 8.40% was observed between moisture levels with the fibers present, with the reduction of voids by approximately 73.34% compared with natural soil.

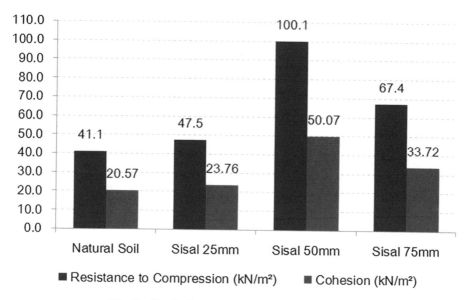

Fig. 5. Simple Compression and Cohesions graphs

Table 4. Preconsolidation pressure of each sample

Sample	Preconsolidation pressure
Natural soil	196.5364
Soil + sisal fiber of 25 mm	148.0256
Soil + sisal fiber of 50 mm	128.6354
Soil + sisal fiber of 75 mm	56.4314

Table 5. Initial void index of each sample

Sample	Initial void index
Natural soil	0.9819
Soil + sisal fiber of 25 mm	0.6251
Soil + sisal fiber of 50 mm	0.5225
Soil + sisal fiber of 75 mm	0.5202

Figure 6 presents graphs of Void Indices × Pressure (kPa) of the samples. It can be seen that, the addition of sisal fiber reduces the void index and increases its plasticity compared with natural soil. It can be seen that 25 mm fiber was the best on reduced the void indices and obtained the greatest increase in plasticity. The samples with fibers of 50 and 75 mm did not have significant changes, but in comparison to the natural soil, the 50 mm fiber had a lower initial value in the void index.

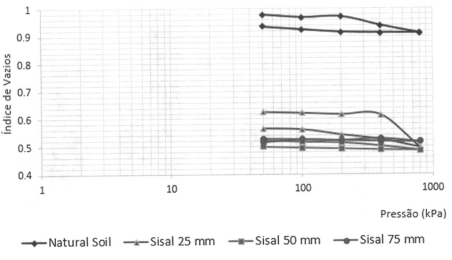

Fig. 6. One-dimensional consolidation test

4   Conclusions

The following conclusions can be drawn from the results and analyses provided in the previous sections. The main conclusions regarding the addition of sisal fibers and their mechanical behavior in reinforced silt-sandy soil are based on the CBR and simple compression tests, as well as the influence of fiber addition on the results of the one-dimensional consolidation test.

1. The addition of 50 mm sisal fibers result in a greater increase in resistance in the CBR test, as well as reduction of the expansion of the analyzed soil. By means of the ISC of natural soil, it can be seen that such soil can be used in sub-grade layers, but is impractical for the sub-layer of pavements. However, with the addition of 50 mm fibers, the CBR provides resistance within the normative standards for application to sub-base layers of paving. The expansions are within the established permissible limits.
2. The soil-fiber mixture using 50 mm long fibers showed greater increase in resistance to simple compression compared with other sisal fiber lengths, and conferred a greater increase in soil cohesion.
3. The increase in sisal fiber lengths in the soil in a unidimensional consolidation test reduced the void indices by approximately 73.34%. The fiber length of 25 mm showed the greatest efficiency in reducing the void index among the analyzed mixtures. The addition of sisal fibers also increased soil plasticity. Thus, the increase of sisal fibers to 25 mm contributes to the reduction of the induced magnitude and the rate of both differential and total settlements of the soil.

The fiber length of 50 mm provided the soil with a greater increase in mechanical properties compared with the other lengths evaluated, reaching a peak in all resistance tests; it was the optimal fiber size found in the present study. For the preconsolidation test, the 25 mm sisal fiber is the most efficient in reducing void indices and increasing soil plasticity. The true efficacy of the use of 75 mm fiber in the soil could not be proven owing to the difficulty of homogenization and high fiber agglomeration during the compaction tests.

The application of 50 mm long sisal fibers at 0.5% content can be used as low-cost geotechnical reinforcement owing to its natural abundance in the Brazilian region.

Acknowledgements. The authors thank God, the university Universidade Católica de Brasília, and Universidade de Brasília for the support provided accomplish this study.

References

American Society for Testing and Materials: Standard Test Method for One-Dimensional Consolidation Properties of Soils. D 2435-96

Associação Brasileira de Normas Técnicas: NBR 6459: solo – determinação do limite de liquidez. Rio de Janeiro, Brazil (1984)

Associação Brasileira de Normas Técnicas: NBR 6508: Grãos de solo que passam na peneira de 4,8 mm - determinação de massa específica. Rio de Janeiro, Brazil (1984)

Associação Brasileira de Normas Técnicas: NBR 7181: Solo – análise granulométrica. Rio de Janeiro, Brazil (1984)

Associação Brasileira de Normas Técnicas: NBR 7182: Solo – ensaio de compactação. Rio de Janeiro, Brazil (1986)

Associação Brasileira de Normas Técnicas: NBR 7180: agregados – determinação do limite de plasticidade. Rio de Janeiro, Brasil (1987)

Associação Brasileira de Normas Técnicas: NBR 9895: Solo – índice de suporte Califórnia. Rio de Janeiro (1987)

Associação Brasileira de Normas Técnicas: NBR 7250/1982 – Identificação e descrição de amostras de solos obtidas em sondagens de simples reconhecimento dos solos. Rio de Janeiro, Brazil (1982)

Bueno, B.S.: Aspectos da estabilização de solos com uso de aditivos químicos e de inclusões plásticas aleatórias. 1996. Concurso Público (Livre Docência) – Universidade de São Paulo, São Carlos (1996)

Departamento Nacional de Estrada e Rodagem: DNER – IE 004: Solos coesivos – determinação da compressão simples. Rio de Janeiro, Brazil (1994)

Departamento Nacional de Infraestrutura de Transporte: Pavimentos flexíveis – Sub-base estabilizada granulometricamente – Especificação de serviço. Rio de Janeiro, Brazil (2009)

Donisete, I.: Ensaio de Índice de Suporte Califórnia – CBR (2016). http://lpe.tempsite.ws/blog/index.php/ensaio-de-indice-de-suporte-california-cbr/. Accessed 03 Nov 2017

Gray, D.H., Ohashi, H.: Mechanics of fiber reinforcement in sand. J. Geotech. Eng. **109**(3), 335–353 (1983)

Machado, S., Machado, M.: Mecânica dos Solos I. 1st. edn., pp. 84–85, Rio de Janeiro (2006)

Murthy, V.N.S.: Soil Mechanics and Foundation Engineering. CBS Publishers & Distributors, New Delhi (2009). 665 p.

Pinto, C.S.: Curso Básico de mecânica dos solos. 3rd edn. Oficina de Textos, São Paulo 77,354 (2006)

Sampaio, J.L.: Estudo da compressibilidade de uma argila mole da região metropolitana de Belém, por meio de ensaios de adensamento. In: 15th Congresso Brasileiro de Mecânica dos Solos e Engenharia Geotécnica, COBRAMSEG, Porto de Galinhas, Brazil (2010)

Trindade, A.: Iniciação científica: Estudo de materiais alternativos para utilização como novos materiais geotécnicos - aplicabilidade de fibras naturais de sisal como reforço de solos, pp. 2–5. PUC, Rio de Janeiro (2016)

Vargas, M.: Introdução à mecânica dos solos. Mcgraw Hill, USP (1977)

Bearing Capacity of Surface Treated Coir Geotextile Reinforced Sand

R. Sridhar[1](✉) and M. T. Prathapkumar[2]

[1] Department of Civil Engineering, Ghousia College of Engineering,
Ramanagara, Karnataka, India
rsridharmanasa@gmail.com
[2] Department of Civil Engineering, RNS Institute of Technology,
Bangalore, Karnataka, India
mtprathap-63@gmail.com

Abstract. Natural fiber used as Reinforcement in composites has in modern times paying attention due to low cost, easy availability, low density, acceptable specific properties, ease of separation, enhanced energy recovery, CO_2 neutrality, biodegradability and recyclable in nature. Although glass and other synthetic fiber-reinforced plastics have high specific strength, their fields of application are very restricted because of their inherent higher cost of production. In this connection, an investigation has been carried out to make use of coir; a natural fiber abundantly available in India. Whenever the water absorption capacity of fiber reduces the longevity will be increased. Coir can be efficiently used as reinforcing material if a proper treatment is adopted. Whenever the water absorption capacity of fiber reduces the longevity will be increased. Natural-fibre-reinforcement when treated with chemicals could be successfully produced with good mechanical properties and the tensile and flexural properties can be further enhanced by alkali treatment. In the present investigation, the coir geotextile is proposed to be used as reinforcement in sand has been made less water absorbent by surface. Chemical treatment of reinforcement material is one of the main ways of improving mechanical properties of natural fiber reinforced polymer composites treatment using Sodium Hydroxide. The effect of surface treated coir mat was used as reinforcement in the form of layer below the model footing, to assess and compare its performance with untreated coir mat reinforced sand. Peak load intensity was evaluated to determine the optimum position of layer for the geotextile reinforcement was determined experimentally.

Keywords: Coir geotextile · Natural fibers · Treated coir · Peak load intensity

1 Introduction

Foundation is that component of the structure which is in direct contact with the soil and transmits load directly to the soil. Prior to the application of load, the soil beneath the base of the footing is in elastic equilibrium. As the load is applied settlement occur which is proportional to the load. With the increase in loading, settlement progressively increases, and the soil transforms from the state of elastic equilibrium to plastic

© Springer Nature Switzerland AG 2019
M. Meguid et al. (Eds.): GeoMEast 2018, SUCI, pp. 25–34, 2019.
https://doi.org/10.1007/978-3-030-01944-0_3

equilibrium and thus the distribution of soil reaction changes and failure of soil occurs. The improved performance of reinforced soil foundation can be attributed to three fundamental reinforcement mechanisms. If the top layer spacing is greater than a certain value, the reinforcement would act as a rigid boundary and the failure would occur above the reinforcement.

Binquet and Lee were the first who reported this finding. Experimental study conducted by several researchers like Sitharam and Sireesh (2004), Madhavi Latha and Somwanshi (2009) confirmed this finding subsequently. Under loading, the footing and soil beneath the footing move downward. As a result, the reinforcement is deformed and tensioned. Due to its tensile stiffness, the curved reinforcement develops an upward force to support the applied load. A certain amount of settlement is needed to mobilize tensioned membrane effect and the reinforcement should have enough length and stiffness to prevent it from failing by pull out and rupture. This mechanism was applied to develop a design method for a strip footing on reinforced sand with the simple assumption made for the shape of reinforcement after deformation and later it was extended to a rectangular footing on reinforced sand. Coir fiber is strong fiber among all natural fibers. Coir is a biodegradable natural fibrous material containing 40% lignin and 54% cellulose (Rao and Balan 2000). Because of its high lignin content, Coir can be used in various forms (Sivakumar Babu and Vasudevan 2008a, b). Many researchers (Sivakumar Babu and Vasudevan 2008a; Sivakumar and Choksey 2010; Binquet and Lee 1975; Dasaka and Sumesh 2011; Mwasha 2009; Dutta et al. 2012; Madhavi Latha and Somwanshi 2009; Prasad et al. 1983; Rao et al. 2005, 2006) have shown that coir fibre reinforcement can significantly improve the engineering properties of soil. But the presence of the pectins, lignin, hemicellulose, silica and pith on the surface of these fibres results in poor interaction with the soil (Prasad et al. 1983). Significant increase in strength parameters and stiffness of sand reinforced with coir fibres was reported by Rao and Balan (2000). The strength and stiffness of tropical soil were increased with the inclusion of discrete coir fibres of about 1–2% by weight (Sivakumar Babu and Vasudevan 2008a, b). The coir fibres have good strength and resistance to bio-degradation over a long period of time (Mwasha 2009; Dutta et al. 2012). The unconfined compressive Strength of black cotton soil reinforced with bitumen coated coir fibres shows marginal variation in strength as compared to uncoated coir fibres (Sivakumar Babu and Choksey 2010). The results of the tests and the model were quite comparable. The unconfined compressive strength test and unconsolidated undrained triaxial test on low compressible clay reinforced with coir fibres were reported by Dasaka and Sumesh (2011). The results of the unconfined compressive strength test with the addition of 30 mm and 15 mm long coir fibres to the clay indicated a decrease and an increase in its unconfined compressive strength, respectively. however it needs further treatment in order to prolong its service life as reinforcing material to soil. In general, coir fiber is a natural organic material, which absorbs huge water content and has a short life period and it is bio-degradable. Natural fiber used as Reinforcement in composites has in recent times attracted attention due to low cost, easy availability, low density, acceptable specific properties, ease of separation, enhanced energy recovery, CO_2 neutrality, biodegradability and recyclable in nature. Although glass and other synthetic fibre-reinforced plastics possess high specific strength, their fields of application are very limited because of their inherent higher cost of production. In this connection, an investigation has been carried out to make use of coir; a natural fiber abundantly available in India. Coir has a longer life compared to other natural fibers that degrade much faster, it is possible to use these fibers in rural roads and ground improvement (Rao et al. 2006). Coir fiber is strong fiber among all natural fibers, however it needs further treatment in order to prolong its service life as reinforcing material to soil. In general, coir fiber is a natural organic material, which absorbs huge water content and has a short life period and it is bio-degradable. Whenever the water absorption capacity of fiber reduces the longevity will be increased. Coir can be efficiently used as reinforcing material if a proper treatment is adopted. Whenever the water absorption capacity of fiber reduces the longevity will be increased. Natural-fibre-reinforcement when treated with chemicals could be successfully produced with good mechanical properties and the tensile and flexural properties can be further enhanced by alkali treatment. In the present investigation, the coir mat is proposed to be used as reinforcement in sand has been made less water absorbent by surface. Chemical treatment of reinforcement material is one of the main ways of improving mechanical properties of natural fiber reinforced polymer composites treatment using Sodium Hydroxide. The effect of surface treated coir mat was used as reinforcement in the form of layer below the model footing, to assess and compare its performance with untreated coir mat reinforced sand. BCR was evaluated to determine the optimum position of layer for the geotextile reinforcement were determined experimentally.

2 Materials Used

2.1 Sand

A naturally occurring granular material composed of finely divided rock and mineral particles was used which was obtained locally in Bengaluru. Properties of sand used are as shown in Table 1.

Table 1. Properties of sand used

Coefficient of uniformity (C_u)	4.48
Coefficient of curvature, C_c	0.960
Specific gravity, G	2.66
Maximum density of sand, Υ_d (max.), kN/m^3	16.7
Minimum density of sand, Υ_d (min.), kN/m^3	14.0
Classification of Sand	SP

2.2 Coir Geotextile

Coir Geotextile was obtained from Karnataka coir federation, Bengaluru, Karnataka, India. The properties of coir geotextile are as shown in Table 2.

Table 2. Properties coir geotextile

Mass/unit area (g/m^2)	835
Thickness (mm)	6.81
Yarn count	
Direction A (Ne)	2/0.24^5
Direction B (Ne)	2/0.22^5
No. of yarns/dm	
Direction A/dm	7
Direction A/dm	9
Yarns twist (Turns/m)	
Direction A	73
Direction B	63
Cover factor	10.8
Breaking load (kgf)	25.2
Elongation (%)	31

2.3 Chemicals

Chemicals like sodium hydroxide, Ethanol, Benzene and water based epoxy resin were used to modify the surface characteristics of coir mat.

2.4 Surface Modification

As per the review of literature (Prasad et al. 1983), alkali treated coir fibers using sodium hydroxide of 4% concentration is found to reduce water absorption of coir fibers by 70%. In addition, to make the coir fiber hydrophobic, it was also coated with water based epoxy resin. Such a coating was found to make the surface of coir fibers smooth. Hence, the epoxy coated coir fibers was sprayed with liberal doses of stone dust powder, which sticks to the surface randomly. This enhances the frictional characteristics of surface of coir, which is essential for interaction between reinforcement and soil. The treatment of coir fibers were done in stages: (i) The fibres were first scoured with detergent solution (2%) at 70 °C for 1 h and washed with distilled water and dried in vacuum oven at 70 °C. (ii) Fibres are then immersed in a 1:2 mixture of Ethanol and Benzene for 72 h to dewax the sample followed by washing with distilled water and air drying. (iii) Next the fibres were treated with NaOH solution for 3 h at 100 °C with occasional shaking followed by washing with distilled water to obtain alkali treated fibres. (iii) Further with these treated fibre materials, fibres of length 20 mm were cut and mats of diameter 28 cm with mat openings of size 10 × 10 mm, 20 × 20 mm and 30 × 30 mm were prepared. (iv) Further these fibres and mats were sprayed with a water based epoxy resin [with 1(base):1(hardner):2(water)],which is available in the market with the name Dr.fixit Dampguard and were air dried for 6–8 h. (v) Epoxy coated fibres were immediately sprinkled with stone dust passing 75μ and were air dried for 6–8 h.

2.5   Stone Dust

Stone dust is a byproduct of cutting or crushing of stone. Stone dust was collected from a stone query, Bangalore. Stone dust passing 75μ sieve was used and the specific gravity of stone dust used is 2.6.

3   Methodology

The load tests were conducted using a load frame. Sand beds were prepared in a cylindrical steel tank of diameter 30 cm and height 35 cm and compacted in layers of 5 cm to have a relative density of 60%. The model footing used for the tests was circular in shape and is of 5 cm diameter (B). Coir geotextile having diameter slightly less than diameter of tank were used with opening 20 mm, placed at specific depths while preparing the sand bed for each model test. The depth of first layer of reinforcement from the bottom of the footing is measured as u, and model footing tests for depth of first location of reinforcement to width of footing ratio was kept equal to 0.3 (u/B). Tests were also conducted using untreated coir geotextile under the different locations of u/B ratios of 0.6,1 and 2 for comparative analysis. The load tests were conducted for treated coir geotextile at different locations in sand as mentioned above for the case of untreated coir geotextile were conducted.

Using load settlement test plots of load intensity in kN/m^2 versus percentage strain were plotted both for untreated and treated coir geotextile used in present investigation.

4   Results and Discussion

4.1   Effect of Location of Coir Geotextile in Sand

Results of the load settlement measurement were plotted in terms of Load-intensity versus Percent Strain for model footings resting on unreinforced and reinforced sand beds. Typical curves at U/B ratio of 0.3, 0.6, 1.0 and 2.0 are as shown in Fig. 1. The peak stress was obtained from these plots and the strain corresponding to the peak stress was considered as peak strain. Load–settlement behaviour of sand showed general shear failure, indicating that the sand was stiffened with the inclusion of reinforcement.

Lifting of coir mat placed at shallow depth ratio of u/B = 0.3 was observed beyond peak stress value measured during loading process. Considerable bulging was observed around the edges of model footiongs were observed for sand reinforced with coir fibers upto a shallow depth ratio of u/B = 0.3 and 0.6.

Figure 1 shows the effect (u/B) ratio on bearing capacity for three differen locations of coir geotextile in foundation. From the test results on footing has been found that load intensity increases during reinforced condition than that of unreinforced condition. The increase in bearing capacity with the addition of geotextile is due to the fact that reinforcing elements interact with the soil particles mechanically through surface tension and by interlocking. The function of the interlock or bond is to transfer the load from soil to the reinforcing element by mobilizing the tensile strength of reinforcing

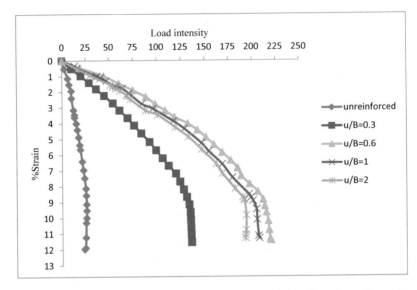

Fig. 1. Load-settlement characteristics of unreinforced and reinforced sand at various u/B ratios.

elements which results into the improvement in bearing capacity. Similar trend has been observed by the earlier researchers. Sridhar et al. (2011) reported that reinforced soil medium gives more bearing capacity than the unreinforced soil bed for square footing and concluded that as reinforcement layer increases the bearing capacity of the soil medium also increases.

Figure 2 shows the results of the load settlement measurement were plotted in terms of Load-intensity versus Percent Strain for model footings resting on and reinforced sand beds for the case of treated coir geotextile.

Comparison between load settlement characteristics of untreated and treated coir geotextile reinforced sand has been shown in Fig. 3. It can be observed that due to the alkali treatment provided to coir geotextile shown improvement with respect to bearing capacity from the range of 8% to 10%. Therefore it can be concluded that treatment of coir geotextile increases the bearing capacity which when compared to untreated coir geotextile is larger. Hence treatment of coir mat is more beneficial in terms of increasing bearing capacity. Treated coir geotextile specimens can be attributed to better interaction of sand with fibre matrix.

The peak load intensity of the sand reinforced with coir geotextile can be significantly improved with the treatment with alkali treatment. Similarly increase in peak load intensity of treated coir geotextile reinforced sand has shown in Figs. 4, 5 and 6 for u/B ratios 0.6, 1 and 2 respectively. To prove better performance of treated coir reinforcement, the Table 3 as been formed. The increase in peak load intensity may observed for different % strain at various u/B ratios for treated coir geotextile reinforced sand in Table 4.

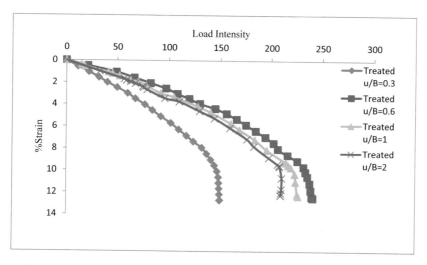

Fig. 2. Load-settlement characteristics of treated coir geotextile reinforced sand at various u/B ratios.

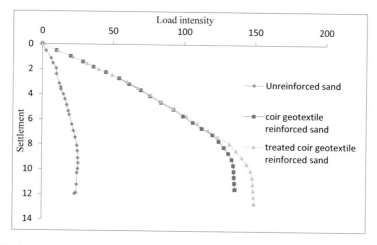

Fig. 3. Variation of load settlement characteristics of treated coir geotextile reinforced sand at u/B = 0.3

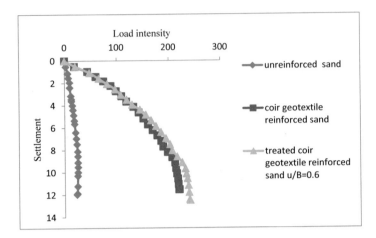

Fig. 4. Variation of load settlement characteristics of treated coir geotextile reinforced sand at u/B = 0.6

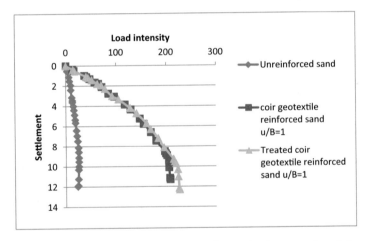

Fig. 5. Variation of load settlement characteristics of treated coir geotextile reinforced sand at u/B = 1

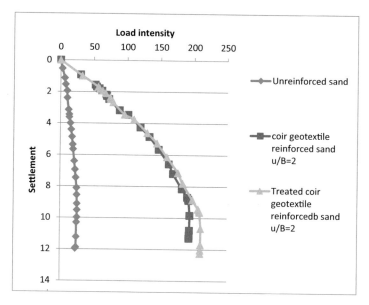

Fig. 6. Variation of load settlement characteristics of treated coir geotextile reinforced sand at u/B = 2

Table 3. Peak load at various %strains for different u/B ratios in untreated coir geotextile reinforced sand

u/B ratio/% Strain	0.3	0.6	1	2
2%	40	77	68	65
4%	74.5	132.5	121	116
6%	103	170	157	160
8%	127	204	195	185

Table 4. Peak load at various %strains for different u/B ratios in treated coir geotextile reinforced sand

u/B ratio/% strain	0.3	0.6	1	2
2%	43.72	83.12	74.43	70.12
4%	80.5	143.45	130.86	125.21
6%	109.3	185.3	169.88	174.23
8%	136.3	218.2	208.33	202.23

5 Conclusions

On the basis of present experimental study, the following conclusions have been drawn:

(i) Coir geotextile when placed in failure zone is the most beneficial in terms of increasing bearing capacity.

(ii) The peak stress of surface treated coir geotextile is larger than untreated coir fibers under a given u/B ratio.

(iii) Surface treatment of coir products make them hydrophobic and a more beneficial by product for soil reinforcement.

(iv) It can be observed that due to the alkali treatment provided to coir geotextile shown improvement with respect to bearing capacity from the range of 8% to 10%. Therefore it can be concluded that Treatment of coir geotextile increases the bearing capacity which when compared to untreated coir geotextile is larger.

References

Sivakumar Babu, G.L., Vasudevan, A.K.: Strength and stiffness response of coir fibre-inforced tropical soil. J. Mater. Civ. Eng. **20**(9), 571–577 (2008a). https://doi.org/10.1061/(ASCE) 0899-1561(2008)20:9(571)

Sivakumar Babu, G.L., Choksey, S.: Model for analysis of fibre-reinforced clayey soil. Geomech. Geoeng. Int. J. **5**(4), 277–285 (2010). https://doi.org/10.1080/17486021003706804

Binquet, J., Lee, K.L.: Bearing capacity tests on reinforced earth slabs. J. Geotech. Eng. Div. ASCE **101**(12), 1241–1255 (1975)

Dasaka, S.M., Sumesh, K.S.: Effect of coir fibre on the stress–strain behavior of a reconstituted fine-grained soil. J. Nat. Fibres **8**(3), 189–204 (2011). https://doi.org/10.1080/15440478. 2011.601597

Mwasha, P.A.: Coir fibre: a sustainable engineering material for the Caribbean environment. Coll. Bahamas Res. J. **15**, 36–44 (2009)

Dutta, R.K., Khatri, V.N., Gayathri, V.: Effect of addition of treated coir fibres on the compression behaviour of clay. Jordan J. Civ. Eng. **6**(4), 476–488 (2012)

Madhavi Latha, G., Somwanshi, A.: Bearing capacity of square footings on geosynthetic reinforced sand. Geotext. Geomembr. 281–294 (2009)

Prasad, S.V., Pavithran, C., Rohatgi, P.K.: Alkali treatment of coir fibres for coir-polyester composites. J. Mater. Sci. **18**(5), 1443–1454 (1983). https://doi.org/10.1007/BF01111964

Rao, G.V., Dutta, R.K., Ujwala, D.: Strength characteristics of sand reinforced with coir fibers and coir geotextiles. Electron. J. Geotech. Eng. **11** (2006)

Rao, G.V., Dutta, R.K., Damarashetty, U.: Strength characteristics of sand reinforced with coir fibres and coir geotextiles. Electron. J. Geotech. Eng. **10**(G) (2005)

Rao, G.V., Balan, K.: Coir Geotextiles - Emerging Trends. Kerala State Coir Corporation Limited, Alappuzha (2000)

Sitharam, T.G., Sireesh, S.: Model studies of embedded circular footing on geogrid-reinforced sand beds. Ground Improv. **8**(2), 69–75 (2004)

Sivakumar Babu, G.L., Vasudevan, A.K.: Seepage velocity and piping resistance of coir fiber mixed soils. J. Irrig. Drain. Eng. **134**(4), 485–492 (2008b)

Load Carrying Capacity of Expansive Clay Beds Reinforced with Geogrid-Encased Granular Pile-Anchors

A. S. S. Raghuram[1]($\boxtimes$), B. R. Phanikumar[2], and A. S. Rao[3]

[1] Department of Civil Engineering, Indian Institute of Technology Hyderabad, Kandi 502285, India
raghuram.ammavajjala@gmail.com
[2] Department of Civil Engineering, S. R. K. R. College of Engineering, Bhimavaram 534204, India
phanikumar_29@yahoo.com
[3] Department of Civil Engineering, UCEK (A), JNTUK, Kakinada 533003, India
sreeramaajjarapu@gmail.com

Abstract. Plate load tests were conducted on unreinforced expansive clay beds and clay beds reinforced with Granular Pile Anchor (GPA) and geogrid-encased GPA to compare their compressive load response. It was found from the tests that the expansive clay beds reinforced with geogrid-encased GPA showed higher load-carrying capacity and improved compressive load response compared with GPA and unreinforced beds. The applied pressure to cause a settlement of 25 mm in the unreinforced clay bed and clay bed reinforced with GPA of length 97 mm and diameter 30 mm ($\frac{l_{gp}}{d_{gp}} = 3.23$) was respectively 152.78 kPa and 789.4 kPa, showing an improvement of 416.7%. However, the clay bed reinforced with geogrid-encased GPA required a pressure of 858.16 kPa for the same amount of settlement. This shows an improvement of 461.7% with respect to the unreinforced clay bed. Among three types of anchors used in this study, the one with a length of 97 mm and a diameter of 30 mm showed the best load response when tested without geogrid, resulting in an improvement of 659.3% whereas, the same GPA showed a better improvement of 955.57% when encased with geogrid.

Notations

GPA	Granular Pile Anchor
l_{gp}	Length of GPA
d_{gp}	Diameter of GPA
PVC	Poly vinyl chloride
G	Specific gravity
D_r	Relative density
γ_d	Maximum dry density
OMC	Optimum moisture content
c'	Cohesion
ϕ'	Internal angular friction

© Springer Nature Switzerland AG 2019
M. Meguid et al. (Eds.): GeoMEast 2018, SUCI, pp. 35–44, 2019.
https://doi.org/10.1007/978-3-030-01944-0_4

δ Interface friction angle
e_o Natural void ratio

1 Introduction

Different foundation techniques have been suggested for mitigating heave of expansive clays. Some of them are sand cushion, the Cohesive Non-Swelling layer (Katti et al. 2002) belled piers, granular pile anchor foundations (Phanikumar 1997; Phanikumar et al. 2004; Rao et al. 2007, 2008; Phanikumar et al. 2008). Granular Pile Anchor (GPA) is an innovative foundation technique, developed for improving the engineering behavior of expansive clay beds. GPA is a modification of the conventional granular pile where an anchor is placed centrally in the pile and connected to the foundation at the top and a base plate at the bottom to render the pile tension.

Encasement of the GPA into a geofabric tube has effected the have and shrinkage of the reactive soils by using finite element analysis (Ismail and Shahin 2011, 2012). The authors have also demonstrated the group effect of the GPA in arresting heave. Pullout resistance of GPA increased by reinforcing GPA with base geosynthetics (Phanikumar and Ramachandra 2000; Phanikumar 2016). They have stated that pullout capacity of the GPA increased by reinforcing the GPA with geogrid layers which are placed at different depths from the base of the GPA. Heave decreased drastically when GPA are encased with geogrid (Raghuram et al. 2016; Johnson and Sandeep 2016; Raghuram and Rao 2017). The heave behavior of expansive clay beds reinforced with encased GPA of varying stiffness (Muthukumar and Shukla 2016). The authors have reported that, the heave of the expansive clay beds was reduced significantly with increasing the stiffness of the geosynthetic. In the present study, geogrid is encased to the GPA, and laboratory scale model tests were conducted on unreinforced expansive clay bed and clay beds reinforced with GPA and geogrid-encased GPA.

2 Experimental Program

A series of laboratory model tests were conducted in this experimental program. The main aim of this investigation was to examine the efficacy of the geogrid-encased GPA. The compressive load response of the composite system was studied by conducting thirteen plate loading tests in CBR moulds.

2.1 Test Materials

The expansive clay used in this study was collected at a depth of 1–1.5 m below the ground surface, from Amalapuram town in the state of AP, India. The index properties of this soil are given in Table 1. The granular material that was used for the installation of the granular piles was a mixture of 20% stone chips with particle size ranged from 6 mm to 10 mm and 80% coarse sand with a size between 2.4 to 4.8 mm. Previous studies showed that this proportion of 20:80 gives the largest difference between the

Table 1. Index properties of expansive soil

Property	Value
Specific Gravity	2.69
Sand (%)	13.33
Silt (%)	35.4
Clay (%)	51.27
Liquid Limit (%)	90
Plastic Limit (%)	35.2
Plasticity Index (%)	54.8
Free Swell Index (%)	140
USCS Classification	CH

maximum and minimum void ratios (Phanikumar 2016). Poly Vinyl Chloride (PVC) coated glass fiber geogrid which had an aperture size 10 mm × 10 mm was overlapped to get an aperture size of 4 mm × 4 mm to be used in the tests. The geogrid has a tensile strength of 10 kN/m and was flexible to enwrap the granular pile-anchor. The tested clay soil had a maximum dry unit weight (γ_d) of 14 kN/m^3 at an optimum moisture content (OMC) of 27.7% as determined by the compaction test. The specific gravity (G) of the granular material was determined by using pycnometer. The maximum and minimum void ratios of the granular material were determined from relative density test (D_r). The natural void ratio (e_o) of the granular material was calculated by choosing the relative density at which granular material was to be compacted. In this study, granular piles were compacted up to a relative density of 70%.

2.2 Test Variables

All the expansive clay beds were compacted to a dry unit weight of 14 kN/m^3 and a placement water content of 15%. These conditions were chosen for convenience of compaction. Further, the compaction of the expansive clay bed was done at a water content much less than the OMC (<27.7%) in order to allow sufficient amount of heave. Tests were performed in CBR moulds of size 150 mm in diameter and 175 mm in height without a collar. The thickness of the expansive clay bed layer was kept constant at 97 mm for all tests. The length of the GPA (l_{gp}) was varied as 97 mm, 73 mm, and 49 mm respectively, but the diameter was kept constant at 30 mm.

2.3 Compaction of Expansive Clay Bed

The clay soil was compacted in five layers, each of which was 20 mm thick. The weight of the soil in each layer to give a dry unit weight of 14 kN/m^3 was calculated. Compaction of the soil was carried out using with a rammer. The process of compaction was continued until the expansive clay bed reached the desired thickness of 97 mm.

As shown in Fig. 1 a sand layer was laid at the bottom of the mould and leveled with a rammer. A circular metal cylinder of diameter 130 mm was inserted into the

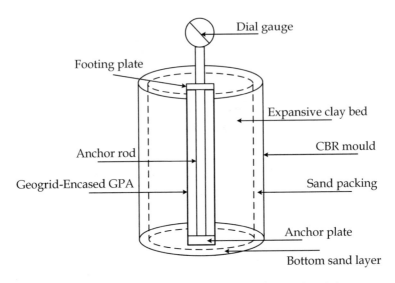

Fig. 1. Experimental set-up showing geogrid-encased GPA foundation system

sand layer so as to form a circular box leaving a space of 10 mm from the sides of the tank. The gap of 10 mm was filled with sand to form sand packing for saturating the expansive clay bed quickly.

2.4 Compaction of Expansive Clay Bed and GPA

The total weight of the soil required to be compacted in the test mould to obtain the pre-determined dry unit weight was calculated. Geogrid of diameter and length equal to those of the GPA was inserted vertically into the bottom sand layer and centered with the CBR mould. A mild steel rod of 5 mm diameter, 30 mm circular steel anchor plate connected to it from the bottom, was vertically inserted into the casing pipe, whereas the anchor plate touching the top surface of the thin sand bedding layer. Moist clay soil was poured into the mould around the casing and compacted with a rammer. Similarly, the weight of granular material required to achieve the pre-determined relative density was calculated. During the process of compaction, the casing pipe and the metal cylinder were withdrawn simultaneously. Sand was poured into this space so that continuity of sand packing was maintained. The process of compaction was continued until the expansive clay bed and granular pile reached the same height. A sand layer, 10 mm thick, was laid over the surface of the compacted expansive clay bed. A surface footing plate was fastened to the top end of the anchor rod.

2.5 Compressive Load Tests

A series of thirteen plate loading tests were conducted to study the stress-settlement behavior for the following cases

Case-I: unreinforced expansive clay bed

Case-II: composite ground (clay with GPA) with and without geogrid. In this case, a footing plate with a diameter greater than the GPS is used to ensure the bearing plate is rested on both the clay soil and the GPA.

Case-III: only GPA with and without geogrid. In this case, a footing plate with a diameter equal to that of the GPA is used

The compressive load was applied after complete saturation of the expansive clay beds. For plate loading tests conducted on Case-I (clay only), a bearing plate of 30 mm was used. In Case-II (composite ground), the footing plate diameter was 50 mm. However, loading GPA alone in Case-III, the diameter of the plate was equal to the diameter of the GPA. The load was applied in increments of $1/5^{th}$ of the expected safe bearing capacity in all the tests. The settlement of the plates under each increment was measured with the help of dial gauges. Tests were continued until one of the two things occurred: (i) either the GPA recorded the settlement of 25 mm, or (ii) the clay soil bed had failed.

2.6   Shear Parameters of Granular Pile-Clay Interface

Shear box tests were conducted to find shear parameters c' and ϕ' for the following cases:

1. Compacting the granular material and expansive clay in the lower and upper halves of the shear box.
2. Compacting the granular material and expansive clay in the lower and upper halves of the shear box with the geogrid in between the granular material and the clay.

These tests were conducted in a shear box of dimensions 60 mm × 60 mm 20 mm to evaluate the friction mobilized at the interface of the clay soil and granular material and interface of clay soil, geogrid and granular material. The values of shear parameters c' and ϕ' were 13 kPa and 30° for the expansive soil-granular pile interface, and 13.3 kPa and 35° for the expansive soil-geogrid-granular pile interface.

3   Discussion of Tests Results

3.1   Compressive Load Response

The plate loading tests were conducted after the attainment of final heave (i.e. at 100% saturation). The attainment of final heave was confirmed when the heave (mm) - time (minutes) plot became asymptotic with the x-axis.

It can be noted from Fig. 2 that the load intensity on the composite ground (Case-II) was more than that on the clay bed alone (Case-I). In Case-II, the applied load would be shared between the saturated clay and the granular pile. For a given settlement of 25 mm, the applied pressure for the unreinforced clay bed was 152.7 kPa whereas for composite ground, in Case-II with a GPA of length 97 mm and diameter

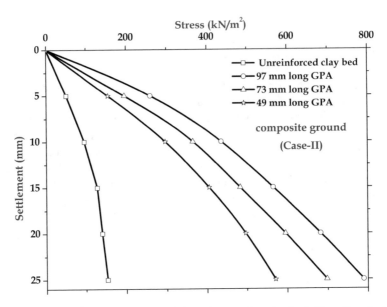

Fig. 2. Stress-settlement behavior of unreinforced expansive clay bed (Case-I) and clay beds reinforced with GPA (Case-II)

30 mm, the applied pressure was 789.4 kPa. Among the various GPAs with three different $\frac{l_{gp}}{d_{gp}}$ ratios (1.63, 2.43 and 3.23), the one with an $\frac{l_{gp}}{d_{gp}}$ ratio of 3.23 (length = 97 mm and diameter = 30 mm) showed the best load response. The other GPAs having lengths of 73 mm and 49 mm respectively resulted in stresses of 697.7 kPa and 570.4 kPa for the same settlement of 25 mm. This can be attributed to the fact that, when the length of GPA is high, a greater amount of resistance would be developed through the interface frictional angle δ.

Figure 3 shows that, for the settlement of 25 mm, the load intensity on the GPA alone (Case-III) was more than that on the unreinforced clay bed. When GPA alone was loaded, the higher load was resisted because of friction angle ϕ' of the granular material. As a result, a higher level of stress was measured in GPA alone compared to the unreinforced clay bed. For instance, the applied pressure for GPA alone of lengths 97 mm, 73 mm and 49 mm was respectively 1160.1 kPa, 1075.3 kPa and 1004.5 kPa for the settlement of 25 mm.

Figure 4 shows that, when geogrid-encased GPA (Case-II with geogrid) was loaded, higher stresses were measured due to geogrid encasement. When geogrid-encased GPA alone (Case-III) was loaded, still higher stresses were recorded compared to the composite ground (Case-II). This is because of the increased interface frictional angle from 30° (soil - granular pile interface) to 35° (soil-geogrid-granular pile interface). As the length of the geogrid-encased GPA increased, the applied pressure to cause the same amount of settlement increased as shown in Figs. 4 and 5.

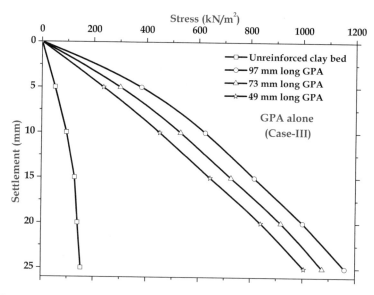

Fig. 3. Stress-settlement behavior of unreinforced expansive clay bed (Case-I) and clay beds reinforced with GPA (Case-III)

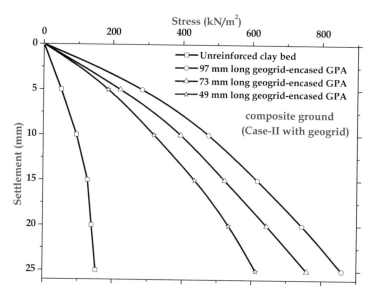

Fig. 4. Stress-settlement behavior of unreinforced expansive clay bed (Case-I) and clay beds reinforced with geogrid-encased GPA (Case-II with geogrid)

The influence of geogrid encasement and type of loading on GPAs is shown in Fig. 6. For a given settlement of 25 mm, the loading intensity required for the unreinforced clay bed was 152.78 kPa whereas, for the composite ground (GPA of length

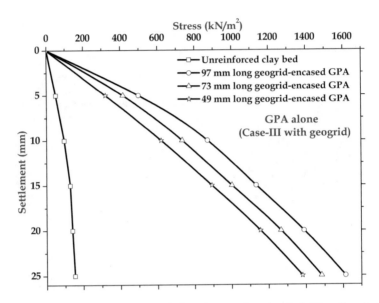

Fig. 5. Stress-settlement behavior of unreinforced expansive clay bed (Case-I) and clay beds reinforced with geogrid-encased GPA (Case-III with geogrid)

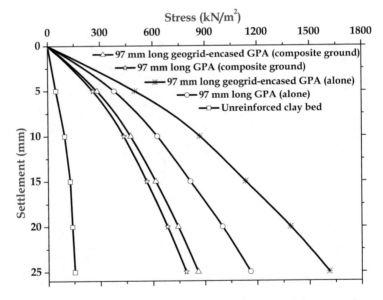

Fig. 6. The influence of geogrid encasement on the behavior of the composite system

97 mm and diameter 30 mm), the load intensity was 789.4 kPa, showing an improvement of 416.7%. However, when geogrid-encased GPA (Case-II) was loaded, the applied pressure was 858.16 kPa, resulting in an improvement of 461.7% with respect to that of the unreinforced expansive clay bed.

Percentage increase in the compressive pressure with reference to the unreinforced clay bed is presented in Fig. 7. For an $\frac{l_{gp}}{d_{gp}}$ ratio of 3.23, the percentage increase in the compressive pressure for the composite ground with respect to the unreinforced clay bed was 416.7% whereas, for the same $\frac{l_{gp}}{d_{gp}}$ ratio of 3.23, the percentage increase in the applied pressure for the geogrid-encased composite ground was 461.7%. And for GPA alone and geogrid-encased GPA alone having the same $\frac{l_{gp}}{d_{gp}}$ ratio, the percentage increase in compressive pressure was 659.3% and 955.57% with respect to the unreinforced clay bed. Moreover, reducing the $\frac{l_{gp}}{d_{gp}}$ ratio resulted in a reduction of the compressive pressure required for both composite ground and GPA alone cases.

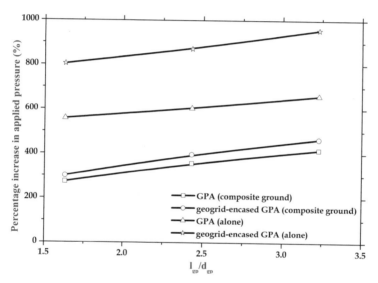

Fig. 7. Percentage increase in applied compressive pressure with respect to unreinforced clay bed

4 Conclusions

The compressive load response of GPAs and geogrid-encased GPAs in expansive clay beds was experimentally investigated. The following are the main conclusions:

1. Using geogrid encasement significantly improved the load carrying capacity of the GPA-soil composite system.
2. Reinforcing the expansive soil with granular pile anchors (with and without geogrid encasement) improved in compressive load response of the clay. For example, the pressure required to cause a settlement of 25 mm was 789.4 kPa and 858.164 kPa respectively for GPA and geogrid-encased GPA of length 97 mm when compared to 152.78 kPa of unreinforced clay.

3. When GPA alone was loaded, the load was resisted by granular material alone. The pressure required to cause the same settlement in GPA alone was 1160.1 kPa which increased to 1612.94 kPa when GPA was encased by geogrid.
4. As the $\frac{l_{gp}}{d_{gp}}$ ratio increased, the pressure required to cause the same settlement increased. This can be attributed to the fact that, when the length of GPA increased, a greater amount of frictional resistance would be mobilized.

References

Ismail, M.A., Shahin, M.A.: Finite element analyses of granular pile anchors as a foundation option for reactive soils. In: Proceedings of International Conference on Advances in Geotechnical Engineering, Perth, Australia, 7–9 November 2011 (2011). ISBN: 978-0-646-55142-5

Ismail, M.A., Shahin, M.A.: Numerical modeling of granular pile-anchor foundations (GPAF) in reactive soils. Int. J. Geotech. Eng. **6**, 149–156 (2012). https://doi.org/10.3328/ijge.2012.06.02.149-156

Johnson, N., Sandeep, M.N.: Ground improvement using granular pile anchor foundation. Procedia Technol. (2016). https://doi.org/10.1016/j.protcy.2016.05.035

Katti, R.K., Katti, D.R., Katti, A.R.: Behaviour of Saturated Expansive Soil and Control Methods. Balkema Publishers, Amsterdam (2002)

Muthukumar, M., Shukla, S.K.: Swelling behaviour of expansive clay beds reinforced with encased granular pile anchors. Int. J. Geotech. Eng. (2016). https://doi.org/10.1080/19386362.2016.1254991

Phanikumar, B.R., Sharma, R.S., Rao, A.S., Madhav, M.R.: Granular pile-anchor foundation system for improving the engineering behaviour of expansive clay beds. Geotech. Test. J. ASTM (2004). https://doi.org/10.1520/GTJ11387

Phanikumar, B.R., Rao, A.S., Suresh, K.: Field behaviour of granular pile-anchors in expansive clays. Ground Improvement J. (2008). https://doi.org/10.1680/grim.2008.161.4.199

Phanikumar, B.R.: A study of swelling characteristics of expansive soils and granular pile-anchor foundation system in expansive soils. Doctoral Thesis, JNTU Hyderabad, India (1997)

Phanikumar, B.R., Ramachandra, Rao N.: Increasing pullout capacity of granular pile anchors in expansive soils using base geosynthetics. Can. Geotech. J. (2000). https://doi.org/10.1139/t00-012

Phanikumar, B.R.: Influence of geogrid reinforcement on pullout response of granular pile-anchors (GPAs) in expansive soils. Indian Geotech. J. (2016). https://doi.org/10.1007/s40098-016-0180-z

Raghuram, A.S.S., Rao, A.S.: Pullout behavior of geogrid-encased granular pile-anchor embedded in expansive clays. In: Proceedings of the 19th International Conference on Soil Mechanics and Geotechnical Engineering, Seoul (2017)

Raghuram, A.S.S., Rao, A.S., Purnanandam, K.: Pullout behaviour of geogrid encased granular pile anchor embedded in expansive clays. In: Proceedings of Indian Geotechnical Conference, Chennai (2016)

Rao, A.S., Phanikumar, B.R., Babu, R.D., Suresh, K.: Pullout behaviour of granular pile-anchors in expansive clay beds in situ. J. Geotech. Geoenviron. Eng. (2007). https://doi.org/10.1061/(ASCE)1090-0241(2007)133:5(531)

Rao, A.S., Phanikumar, B.R., Suresh, K.: Response of granular pile anchors under compression. Ground Improv. J. (2008). https://doi.org/10.1680/grim.2008.161.3.121

Ultimate Bearing Capacity Prediction of Eccentrically Loaded Rectangular Foundation on Reinforced Sand by ANN

R. Sahu[1(✉)], C. R. Patra[1], K. Sobhan[2], and B. M. Das[3]

[1] National Institute of Technology, Rourkela, India
roma.sahu.civ@gmail.com, crpatral9@yahoo.co.in
[2] Florida Atlantic University, Boca Raton, FL, USA
ksobhan@fau.edu
[3] California State University, Sacramento, USA
brajamdas@gmail.com

Abstract. Laboratory model tests were conducted on a rectangular surface foundation resting over multilayered geogrid-reinforced dry sand bed subjected to eccentric load. Based on the model test results, a neural network model was developed to predict the reduction factor that can be used in computing the ultimate bearing capacity of an eccentrically loaded rectangular foundation. This reduction factor (R_k) is the ratio of the ultimate bearing capacity of the foundation subjected to an eccentric load to the ultimate bearing capacity of the foundation subjected to a centric load. A thorough sensitivity analysis was carried out to evaluate the parameters affecting the reduction factor. Based on the weights of the developed neural network model, a neural interpretation diagram is developed to find out whether the input parameters have direct or inverse effect to the output. An ANN equation is developed based on trained weights of the neural network model. The results from artificial neural network (ANN) were compared with the laboratory model test results and these results are in good agreement.

Keywords: Eccentric load · Geogrid · Sand · Neural network
Ultimate bearing capacity · Reduction factor

1 Introduction

During the last three decades, the results of number of studies have been published as related to the ultimate bearing capacity of shallow foundations supported by multi-layered geogrid-reinforced sand and clay. Most of the experimental studies were related to centric loading condition. None of the published studies, however, address the effect of load eccentricity on the ultimate bearing capacity of rectangular surface foundation resting over multi-layered geogrid reinforced sand. The purpose of this paper is to develop a neural network model from the results of laboratory model tests conducted by Sahu et al. (2016) to estimate the reduction factor, R_k. The concept of reduction factor (R_k) is the ratio of the ultimate bearing capacity of the eccentrically loaded rectangular foundation at a depth of reinforcement layer to the ultimate bearing

© Springer Nature Switzerland AG 2019
M. Meguid et al. (Eds.): GeoMEast 2018, SUCI, pp. 45–58, 2019.
https://doi.org/10.1007/978-3-030-01944-0_5

capacity of the vertically loaded foundation at the same depth of reinforcement layer. In the present study, feedforward backpropagation neural network is trained with Levenberg-Marquadrt algorithm. A thorough sensitivity analysis is made to interpret the important input variables. Neural Interpretation Diagram is constructed to find out the direct or inverse effect of input parameters on the output. A prediction model equation is developed based on the weights of the ANN model. Furthermore, the developed reduction factor is compared with the empirical equation proposed by Sahu et al. (2016).

2 Database and Preprocessing

The extensive database obtained from laboratory experiments available in Sahu et al. (2016) have been considered in the present study. Load tests were carried out on model rectangular foundation resting on geogrid reinforced sand subjected to eccentric loads as shown in Fig. 1. The details of the tests and the procedure have been described in Sahu et al. (2016). The database used in the present analysis is presented in Table 1. The database consists of parameters like load eccentricity (e), depth of reinforcement measured from bottom of the foundation (d), width to length ratio (B/L), and ultimate bearing capacity (q_{ue}). Forty-eight numbers of laboratory test results have been taken in this analysis. B/L, d/B and e/B are used as three dimensionless input parameters in the ANN model and the output is the reduction factor (R_k). The reduction factor (R_k) is given by

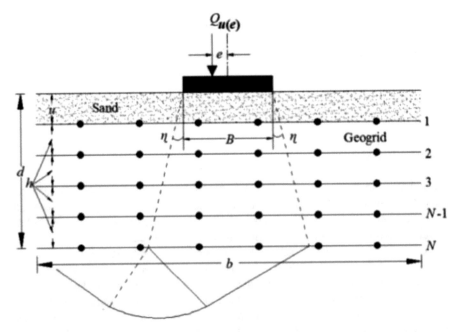

Fig. 1. Eccentrically loaded rectangular foundation on geogrid-reinforced soil

$$R_k = 1 - \frac{q_{uR(B/L,d/B,e/B)}}{q_{uR(B/L,d/B,e/B=0)}} \qquad (1)$$

where $q_{uR(B/L,d/B,e/B)}$ is the ultimate bearing capacity with width to length ratio (B/L) and eccentricity ratio e/B at a depth of reinforcement layer ratio d/B, and $q_{uR(B/L,d/B,e/B=0)}$ is the ultimate bearing capacity with width to length ratio (B/L) and centric vertical loading i.e. $e/B = 0$ at the same depth of reinforcement layer ratio (d/B). Out of 48 data set, 36 data were used for training and 12 data were used for testing which is given in Table 1. Each record represents a complete model test where an eccentrically loaded rectangular foundation is subjected to failure. All the inputs and

Table 1. Database used for ANN model and comparison with Sahu et al. (2016)

Data type (1)	Expt. No. (2)	B/L (3)	d/B (4)	e/B (5)	Experimental q_u (kN/m²) (6)	Experimental R_k (7)	R_k (ANN) (8)	Predicted R_k (9)
Training	1	0	0.6	0.05	337	0.089	0.093	0.1
	2	0	0.6	0.1	290	0.216	0.218	0.22
	3	0	0.6	0.15	240	0.351	0.350	0.35
	4	0	0.85	0	550	0	−0.004	0
	5	0	0.85	0.05	495	0.1	0.101	0.12
	6	0	0.85	0.1	404	0.265	0.263	0.26
	7	0	1.1	0	640	0	0.001	0
	8	0	1.1	0.05	559	0.127	0.121	0.13
	9	0	1.1	0.15	350	0.453	0.465	0.46
	10	0.33	0.6	0	339	0.00	0.00	0.00
	11	0.33	0.6	0.1	267	0.212	0.218	0.22
	12	0.33	0.6	0.15	217	0.360	0.35	0.35
	13	0.33	0.85	0.05	451	0.100	0.106	0.12
	14	0.33	0.85	0.1	371	0.259	0.263	0.26
	15	0.33	0.85	0.15	311	0.379	0.408	0.41
	16	0.33	1.1	0	598	0.000	0.001	0.000
	17	0.33	1.1	0.05	517	0.135	0.141	0.14
	18	0.33	1.1	0.1	416	0.304	0.316	0.3
	19	0.5	0.6	0.00	335	0.000	0.000	0.000
	20	0.5	0.6	0.05	300	0.104	0.096	0.1
	21	0.5	0.6	0.15	213	0.364	0.351	0.34
	22	0.5	0.85	0.00	486	0.000	−0.003	0.000
	23	0.5	0.85	0.1	352	0.276	0.264	0.26
	24	0.5	0.85	0.15	284	0.416	0.408	0.40
	25	0.5	1.1	0.05	493	0.160	0.156	0.14
	26	0.5	1.1	0.1	387	0.341	0.316	0.30
	27	0.5	1.1	0.15	297	0.494	0.465	0.46
	28	1	0.6	0.00	327	0.000	0.002	0.000
	29	1	0.6	0.05	293	0.104	0.101	0.11
	30	1	0.6	0.1	257	0.214	0.218	0.21
	31	1	0.85	0	484	0.000	−0.002	0.00

(continued)

Table 1. (*continued*)

Data type (1)	Expt. No. (2)	B/L (3)	d/B (4)	e/B (5)	Experimental q_u (kN/m^2) (6)	Experimental R_k (7)	R_k (ANN) (8)	Predicted R_k (9)
	32	1	0.85	0.05	421	0.130	0.130	0.13
	33	1	0.85	0.15	292	0.397	0.409	0.40
	34	1	1.1	0	574	0.000	0.002	0.00
	35	1	1.1	0.1	398	0.307	0.316	0.31
	36	1	1.1	0.15	308	0.463	0.465	0.46
Testing	37	0	0.6	0	370	0.000	0.000	0.000
	38	0	0.85	0.15	322	0.415	0.408	0.41
	39	0	1.1	0.1	453	0.292	0.316	0.29
	40	0.33	0.6	0.05	308	0.091	0.095	0.10
	41	0.33	0.85	0	501	0.000	−0.003	0.000
	42	0.33	1.1	0.15	328	0.452	0.465	0.47
	43	0.5	0.6	0.1	259	0.227	0.218	0.22
	44	0.5	0.85	0.05	428	0.119	0.110	0.12
	45	0.5	1.1	0	587	0.000	0.002	0.00
	46	1	0.6	0.15	222	0.321	0.351	0.32
	47	1	0.85	0.1	357	0.262	0.264	0.26
	48	1	1.1	0.05	487	0.152	0.233	0.15

output are normalized in the range of [−1, 1] before training. A feed-forward back-propagation neural network is used with hyperbolic tangent sigmoid function and linear function as the transfer function. The network is trained with Levenberg-Marquardt (LM) algorithm as it is efficient in comparison to gradient descent back-propagation algorithm. The ANN has been implemented using MATLAB V 7.11.0 (R2015b).

3   Results and Discussion

Three inputs and one output parameters were considered in the ANN model. The maximum, minimum, average and standard deviation values of the three input and one output parameters used in the ANN model are presented in Table 2.

The schematic diagram of the ANN architecture is shown in Fig. 2. The number of neurons in hidden layer is varied and it was selected based on the mean square error (MSE) value which was 0.001. In this ANN model three neurons are evaluated in hidden layer as shown in Fig. 3. Therefore the final ANN architecture is retained as 3-3-1 [i.e. 3 (input) – 3 (hidden layer neuron) – 1 (output)].

The coefficient of efficiency (R^2) for the training and testing data are found to be 0.996 and 0.976 respectively which are shown in Figs. 4 and 5. All the data used in the training and testing have been obtained from laboratory model tests and are from the same source and are of same nature. Probably, this may be one of the causes for better fitting in both training and testing phase as well. The weights and biases of the network are presented in Table 3. These weights and biases can be utilized for interpretation of

Table 2. Statistical values of parameters

Parameter	Maximum value	Minimum value	Average value	Standard deviation
B/L	1.0	0	0.457	0.361
d/B	1.1	0.6	0.85	0.204
e/B	0.15	0	0.075	0.055
R_k	0.494	0	0.197	0.156

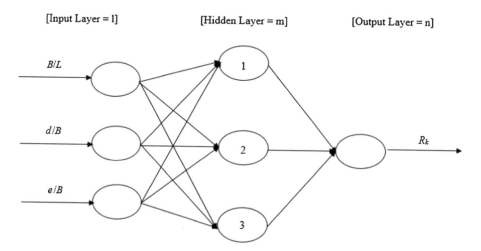

Fig. 2. The ANN architecture

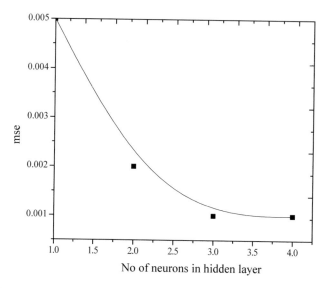

Fig. 3. Variation of hidden layer neuron with mean square error (mse)

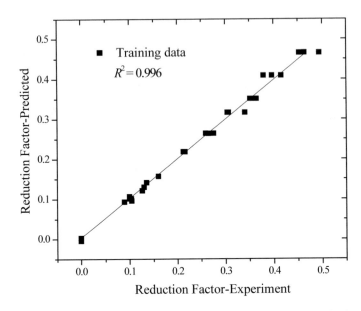

Fig. 4. Correlation between predicted reduction factor with experimental reduction factor for training data

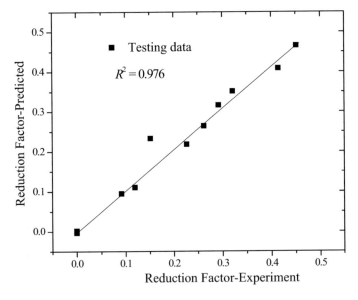

Fig. 5. Correlation between predicted reduction factor with experimental reduction factor for testing data

relationship between the inputs and output, sensitivity analysis and framing an ANN model in the form of an equation. The residual analysis was carried out by calculating the residuals in between experimental reduction factor and predicted reduction factor for training data. Residuals can be defined as the difference between the experimental and predicted R_k value and is given by

$$e_r = (R_k)_i - (R_k)_p \tag{2}$$

Where $(R_k)_i$ and $(R_k)_p$ are the experimental and predicted values of R_k respectively

The residuals are plotted with the experimental number as shown in Fig. 6. It is observed that the residuals are evenly distributed along the horizontal axis of the plot. Therefore it can be said that the network is well trained and can be used for prediction with reasonable accuracy.

Table 3. Connection weights and biases

Neuron	Weight				Bias	
	w_{ik}		w_k		b_{hk}	b_0
	(B/L)	(d/B)	(e/B)	RF		
Hidden neuron 1 ($k = 1$)	0.0006	0.1799	0.7041	1.3085	−0.6441	0.2889
Hidden neuron 2 ($k = 2$)	−0.4876	−0.8140	12.3661	−1.3885	6.1580	
Hidden neuron 3 ($k = 3$)	0.0151	−0.2132	4.2886	1.6768	2.6806	

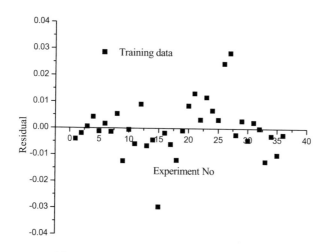

Fig. 6. Residual distribution of training data

4 Sensitivity Analysis

Sensitivity analysis was carried out for selection of important input variables. Different approaches have been suggested in the past to select the important input variables. Connection weight approach by Olden et al. (2004), Garson's algorithm approach by Garson (1991), Pearson correlation coefficient approach by Guyon and Elisseeff (2003) have been applied for sensitivity analysis. The Pearson correlation coefficient is one of them in selecting proper inputs for the ANN model. Goh (1994), Shahin et al. (2002), Behera et al. (2013), Sahu et al. (2017), Sahu et al. (2018), Sethy et al. (2017) and Sethy et al. (2018) have used Garson's algorithm (Garson 1991) in which the input-hidden and hidden-output weights of trained ANN model are partitioned and the absolute values of weights are taken to select the important input variables. It does not provide information on the effect of input variables in terms of direct or inverse relation to the output. Olden et al. (2004) proposed a connection weights approach based on the NID, in which the actual values of input-hidden and hidden-output weights are taken. Table 4 shows the cross-correlation of inputs with the reduction factor. From the table, it is observed that $R_{k(expt)}$ is highly correlated to e/B with a cross-correlation values of 0.975 followed by $d/B = 0.167$ and $B/L = 0.012$. The sensitivity analysis for the model as per Garson's algorithm is presented in Table 5. The e/B is found to be most important input parameter with the relative importance value being 85.39% followed by 10.33% for d/B and 1.32% for B/L. The relative importance of the input variables as calculated following connection weight approach (Olden *et al.* 2004) is also presented in Table 5. As per connection weight approach method e/B is most important input parameter (S_i value −9.05) followed by d/B (S_i value 1.00) and B/L (S_i value 0.70). The S_i value being negative imply that e/B are indirectly and d/B, B/L is directly related to $R_{k(expt)}$ values.

Table 4. Cross-correlation of the input and output for the reduction factor

Parameters	B/L	d/B	e/B	R_{kexpt}
B/L	1	0	0	0.012
d/B		1	0	0.167
e/B			1	−0.975
RF_{expt}				1

Table 5. Relative importance of different inputs as per Garson's algorithm and Connection weight approach

Parameters (1)	Garson's algorithm		Connection weight approach	
	Relative importance (%) (2)	Ranking of inputs as per relative importance (3)	S_i values as per connection weight approach (4)	Ranking of inputs as per relative importance (5)
B/L	1.32	3	0.70	3
d/B	10.33	2	1.00	2
e/B	85.39	1	−9.05	1

5   Neural Interpretation Diagram (NID)

Ozesmi and Ozesmi (1999) proposed neural interpretation diagram (NID) for visual interpretation of the connection weight among the neurons. For the present study, the weights obtained are shown in Table 3. The neural interpretation diagram is presented in Fig. 7. It can be seen from Fig. 7 and 4th column of Table 5 that e/B is inversely related to $R_{k(expt)}$ whereas d/B and B/L is directly related to $R_{k(expt)}$. It can be concluded that $R_{k(expt)}$ value decreases with increase in e/B, but increases with increase in d/B and B/L. In other words, the developed ANN model is not a "black box" and could explain the physical effect of inputs on the output.

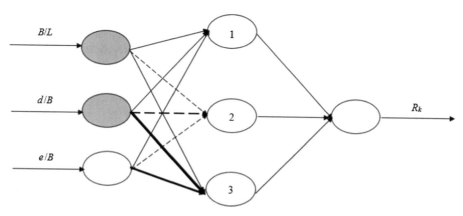

Fig. 7. Neural interpretation diagram showing lines representing connection weights and effects of inputs on reduction factor (R_k)

6   ANN Model Equation for the Reduction Factor Based on the Trained Neural Network

In the present study, with only three parameters (B/L, d/B and e/B) a model equation is developed using the weights obtained from trained neural network model (Goh et al. 2005). The mathematical equation relating the input variables (B/L, d/B and e/B) and the output reduction factor (R_k) can be written as,

$$(R_k)_n = f_n \left\{ b_0 + \sum_{k=1}^{h} \left[w_k f_n \left(b_{hk} + \sum_{i=1}^{m} w_{ik} X_i \right) \right] \right\} \qquad (3)$$

where, $(R_k)_n$ = normalized value of R_k in the range $[-1, 1]$, f_n = transfer function, h = number of neurons in the hidden layer, X_i = normalized value of inputs in the range $[-1, 1]$, m = no. of input variables, w_{ik} = connection weight between i^{th} layer of input and k^{th} neuron of hidden layer, w_k = connection weight between k^{th} neuron of hidden layer and single output neuron, b_{hk} = bias at the kth neuron of hidden layer, b_0 = bias at the output layer.

Thus the model equation for the reduction factor of a shallow rectangular foundation resting over geogrid reinforced sand subjected to eccentric load is formulated using the values of the weights and biases shown in Table 3 as per the following steps.

Step – 1

The input parameters are normalized in the range [−1, 1] by the following expression

$$X_n = 2 \left(\frac{X_1 - X_{min}}{X_{max} - X_{min}} \right) - 1 \tag{4}$$

where, X_n = Normalized value of input parameters, X_{max} = maximum values of the input parameter, X_{min} = Minimum values of the input parameter, X_1 = the data set.

Step – 2

Using the values of the weights and biases presented in Table 3, the following expressions can be written to finally arrive at a correlation of reduction factor $(R_k)_n$ with the input parameters using the following expressions

$$A_1 = 0.0006 \left(\frac{B}{L} \right)_n + 0.1799 \left(\frac{d}{B} \right)_n + 0.7041 \left(\frac{e}{B} \right)_n - 0.6441 \tag{5}$$

$$A_2 = -0.487 \left(\frac{B}{L} \right)_n - 0.814 \left(\frac{d}{B} \right)_n + 12.366 \left(\frac{e}{B} \right)_n + 6.158 \tag{6}$$

$$A_3 = 0.015 \left(\frac{B}{L} \right)_n - 0.213 \left(\frac{d}{B} \right)_n + 4.288 \left(\frac{e}{B} \right)_n + 2.68 \tag{7}$$

$$B_1 = 1.308 \left(\frac{e^{A_1} - e^{-A_1}}{e^{A_1} + e^{-A_1}} \right) \tag{8}$$

$$B_2 = 1.388 \left(\frac{e^{A_2} - e^{-A_2}}{e^{A_2} + e^{-A_2}} \right) \tag{9}$$

$$B_3 = 1.676 \left(\frac{e^{A_3} - e^{-A_3}}{e^{A_3} + e^{-A_3}} \right) \tag{10}$$

$$C_1 = 0.288 + B_1 + B_2 + B_3 \tag{11}$$

$$Rk_n = C_1 \tag{12}$$

Step – 3
Denormalize the $(R_k)_n$ value obtained from Eq. 12 to actual R_k as

$$R_k = 0.5(Rk_n + 1)\,(Rk_{max} - Rk_{min}) + Rk_{min} \tag{13}$$

$$R_k = 0.5(Rk_n + 1)\,(0.494 - 0) + 0 \tag{14}$$

where, Rk_{max} = maximum value of R_k in the database and Rk_{min} = minimum value of R_k in the database.

7 Comparison with Developed Empirical Equation

An empirical equation is developed for reduction factors to predict the ultimate bearing capacity of eccentrically loaded rectangular foundation on reinforced sand for various B/L ratios and expressed as follows:

$$R_k = \frac{q_{uR(e/B,d/B)}}{q_{uR(e/B=0,d/B)}} \tag{15}$$

Thus the reduction factor for various B/L ratio is given as follows

$$R_k = 3.99 \left(\frac{d}{B}\right)^{0.43} \left(\frac{e}{B}\right)^{1.16} \quad \left(\text{for } \frac{B}{L} = 0\right) \tag{16}$$

$$R_k = 3.66 \left(\frac{d}{B}\right)^{0.45} \left(\frac{e}{B}\right)^{1.12} \quad \left(\text{for } \frac{B}{L} = 0.33\right) \tag{17}$$

$$R_k = 3.5 \left(\frac{d}{B}\right)^{0.57} \left(\frac{e}{B}\right)^{1.05} \quad \left(\text{for } \frac{B}{L} = 0.5\right) \tag{18}$$

$$R_k = 2.97 \left(\frac{d}{B}\right)^{0.6} \left(\frac{e}{B}\right)^{1.01} \quad \left(\text{for } \frac{B}{L} = 1\right) \tag{19}$$

As seen in Figs. 8, 9 and Table 1, the comparison seems to be reasonably good. Hence, artificial neural network can be effectively used for the prediction of ultimate bearing capacity of rectangular foundation in geogrid reinforced soil for different B/L ratio under eccentric loading.

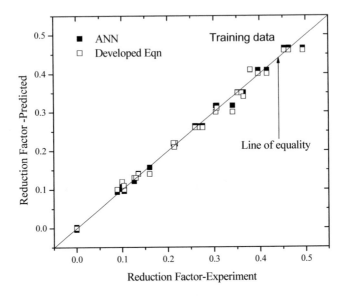

Fig. 8. Comparison of ANN results with Experimental R_k for training data

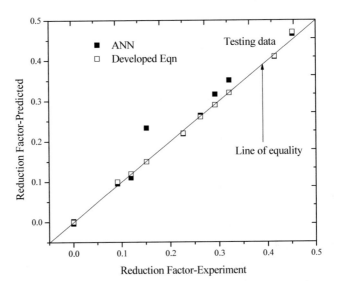

Fig. 9. Comparison of ANN results with Experimental R_k for testing data

8   Conclusions

The following conclusions can be drawn from the above studies

- The result is well trained and can predict the result with reasonable accuracy.
- From ANN model, It is observed that e/B is inversely related to R_k values, whereas, d/B and B/L are directly related to R_k.
- Using Garson's algorithm and Pearson correlation coefficient, e/B is found to be most important input parameter followed by d/B and B/L.
- As per connection weight approach, d/B is found to be most important input parameter followed by e/B and B/L.
- The developed ANN model could explain the physical effect of inputs on the output, as depicted in NID. It is observed that e/B is inversely related to R_k values, whereas, B/L and d/B is directly related to R_k.
- An equation is developed based on the trained weights of the ANN to predict the reduction factor.
- The results obtained from empirical modeling are well agreement with ANN model.

References

Behera, R.N., et al.: Prediction of ultimate bearing capacity of eccentrically inclined loaded strip footing by ANN part 1. Int. J. Geotech. Eng. (2013). https://doi.org/10.1179/1938636212Z.00000000012

Garson, G.D.: Interpreting neural-network connection weights. Artif. Intell. Expert **6**(7), 47–51 (1991)

Goh, A.T.C., Kulhawy, F.H., Chua, C.G.: Bayesian neural network analysis of undrained side resistance of drilled shafts. J. Geotech. Geoenviron. Eng. **131**(1), 84–93 (2005)

Goh, A.T.C.: Seismic liquefaction potential assessed by neural networks. J. Geotech. Eng. ASCE **120**(9), 1467–1480 (1994)

Guyon, I., Elisseeff, A.: An introduction to variable and feature selection. J. Mach. Learn. Res. **3**, 1157–1182 (2003)

Ozesmi, S.L., Ozesmi, U.: An artificial neural network approach to spatial modeling with inter specific interactions. Ecol. Model. (1999). https://doi.org/10.1016/S0304-3800(98)00149-5

Olden, J.D., et al.: An accurate comparison of methods for quantifying variable importance in artificial neural networks using simulated data. Ecol. Model. **178**(3), 389–397 (2004)

Shahin, M.A., et al.: Predicting settlement of shallow foundations using neural network. J. Geotech. Geoenviron. Eng. (2002). https://doi.org/10.1061/(ASCE)10900241(2002)128:9 (785)

Sahu, R., et al.: Ultimate bearing capacity of rectangular foundation on geogrid-reinforced sand under eccentric load. Int. J. Geotech. Eng. (2016). https://doi.org/10.1179/1939787915y.0000000008

Sahu, R., et al.: Use of ANN and neuro fuzzy model to predict bearing capacity factor of strip footing resting on geogrid reinforced sand and subjected to inclined loading. Int. J. Geosynth. Ground Eng. (2017). https://doi.org/10.1007/s40891-017-0102-x

Sahu, R., et al.: Bearing capacity prediction of inclined loaded strip footing on reinforced sand by ANN. Adv. Reinforced Soil Struct. Sustain. Civ. Infrastruct. (2018). https://doi.org/10.1007/978-3-319-63570-5_9

Sethy, B.P., et al.: Application of ANN and ANFIS for predicting the ultimate bearing capacity of eccentrically loaded rectangular foundations. Int. J. Geosynth. Ground Eng. (2017). https://doi.org/10.1007/s40891-017-0112-8

Sethy, B.P., et al.: Prediction of ultimate bearing capacity of eccentrically loaded rectangular foundations using ANN. Adv. Reinforced Soil Struct. Sustain. Civ. Infrastruct. (2018). https://doi.org/10.1007/978-3-319-63570-5_13

Rehabilitation of Canals with Watertight Geomembranes, in the Dry and Underwater

Alberto Scuero[1(✉)] and Gabriella Vaschetti[2(✉)]

[1] Carpi Group, Balerna, Switzerland
alberto.scuero@carpitech.com
[2] Carpi Tech, Balerna, Switzerland
gabriella.vaschetti@carpitech.com

Abstract. Geomembranes are watertight synthetic materials, factory-manufactured under computer-controlled procedures that guarantee constant properties and quality, and supplied as flexible sheets a few millimetres thick. Geomembranes have been used to waterproof all types of hydraulic structures since more than half a century. In canals, their use started after World War II. They have been adopted to restore watertightness of deteriorated concrete or bituminous concrete linings, and to repair failing joints. Besides this water-proofing function, exposed geomembrane systems provide the additional function of increasing the water flow. All types of canals have been lined, be it hydropower canals with high water velocity, or large irrigation canals with varying sections, or small flumes and aqueducts. The paper discusses, through some significant case histories, the design options for the various projects, addressing the anchorage and drainage systems, taking into account the dynamic action of water flowing inside the canal, of water acting behind the geomembrane because present in the slopes, and the generally worst-case scenario of water infiltrated via accidental damage under the geomembrane; the influence of the drainage system and of the type of geomembrane on the anchorage will also be addressed. The paper finally introduces a totally innovative technology, SIBELONMAT®, which allows repairing canals underwater, without impacting on the operation, if needed with un-manned procedures, using ROVs. This new technology, which uses panels of a double geomembrane mattress filled with mortar, and joined with a watertight device, has already been successfully installed underwater in two pilot projects in two water supply canals, with no reduction of water speed. The technology can be considered also for new construction of canals and of embankment dams.

1 Introduction

Canals transport water from the source to users who are generally far away. With the increasing demand in water supply, either for power generation or for irrigation and human consumption, efficient transport of water is a very important issue. Canals must be designed, built and maintained to ensure that the maximum possible amount of water is safely conveyed to the final users at the lowest possible costs. In modern times canals have been lined with low permeability materials such as compacted clay, brick or rock masonry, concrete, and bituminous concrete. As canals age, these linings

© Springer Nature Switzerland AG 2019
M. Meguid et al. (Eds.): GeoMEast 2018, SUCI, pp. 59–76, 2019.
https://doi.org/10.1007/978-3-030-01944-0_6

deteriorate due to the dynamic action of water and of transported sediments, to temperature gradients, to uplift pressure, to settlements in the natural ground, to growth of vegetation, to animals' hoofs, to action of ice and frost, to chemical attack of water, etc. Water is lost through cracks and failing joints, permeability and hydraulic roughness increase, vegetation grows. The canal no more transports the amount of water which it was designed for; in some cases, almost all water cannot be distributed to the users. Water seepage into the surrounding soil may jeopardise the stability and safe operation of the canal, and ultimately trigger landslides.

Watertightness can be restored by local repairs, effective only in case of limited deterioration and in the short term, or by providing a new liner to the canal. Reconstruction of a traditional liner requires long outage, is expensive, and will soon be subject to the same deterioration mechanism. Soon after World War II alternative solutions, using as water barrier the new synthetic watertight materials that had come available to the market, were explored and experimented. The evident technical, construction and economic advantages of the pioneer installations opened the way to the development of more sophisticated solutions, which eventually provided the owners with new options that could efficiently substitute traditional liners.

The materials used as water barriers are flexible thermoplastic geomembranes, supplied in form of factory-manufactured sheets a few millimetres thick, having permeability ($<10^{-6}$ m^3/m^2/day as per EN 14150 standard) one to two or more orders of magnitude lower than that of traditional liners. If the appropriate material is selected, the geomembrane can be installed on fairly uneven surfaces, is able to bridge joints and even large fissures, and to withstand settlements that would destroy a traditional rigid liner. The smooth geomembrane surface, with its very low hydraulic roughness, increases the efficiency of water transport: increase in water flow up to 90% have been documented, as further discussed in the paper.

The geomembrane liner is generally installed on the entire surface of the canal, over the deteriorated old liner, to which it is anchored with a mechanical system that constructs a drainage gap between the geomembrane liner itself and the existing canal surface. Water coming from the inside of the structure (accidental seepage), or migrating towards the canal from the surrounding soil, can thus be drained, collected and discharged by means of the drainage system, avoiding unbalanced back-pressure acting on the geomembrane, and providing the additional benefit of capturing water seeping from the ground with beneficial effects on the stability of the slopes.

2 Design

From the empirical approach adopted in pioneer projects after the end of World War II, application of geomembrane systems evolved to a more scientific approach based on analytical study and testing. After more than 60 years of field experience, there is now extensive literature on the design, installation and performance of geomembrane systems in hydraulic structures. Studies are continuing to improve materials' characteristics and design, to understand and predict at best their field behaviour, and to disseminate knowledge. Among the entities involved since decades in this process are ICID - International Commission on Irrigation and Drainage (Plusquellec 2004),

ICOLD - International Commission on Large Dams (ICOLD 2011), IGS - International Geosynthetics Society, BuRec - US Bureau of Reclamation, USACE - US Army Corps of Engineers (Christensen et al. 1995), and IREQ - Research Institute of the Canadian Hydro Québec (Durand and Tremblay 1995). The most recent substantial contribution in the field of canals was made by the Hydraulic Engineering and Water Management Department of TUM, the Technical University of Munich.

2.1 TUM Research

The issue of fastening systems in canals is a critical issue, because there have been many cases in which insufficient anchorage has caused failure of the geomembrane system. No independent analytical studies were available on this subject, and the need for guidelines to allow appropriate comparison and evaluation of geomembrane sealing systems in the tendering process, and avoid future inconveniences, was deeply felt. TUM carried out a 4-year research program (Strobl and Schaefer 2005; Schaefer 2006) that thoroughly investigated the related issues, i.e. the loads acting on geomembrane liners in canals, how different types of geomembranes transfer such loads to the fastening system, the influence of the drainage system, sealing at boundaries, and calculation of the fastening system in worst-case scenarios. Extensive real scale testing needed to verify performance in the field was performed with a special laboratory equipment to investigate how loads are transferred to the anchorage system, and in the canal of TUM Laboratory of Hydraulics at Obernach to investigate the behaviour of fully equipped geomembrane systems.

The TUM research focused on two types of geomembranes, SIBELON®, a Polyvinylchloride (PVC) composite geomembrane (geocomposite), and a High-Density Polyethylene (HDPE) geomembrane, as representatives respectively of flexible, low modulus materials, and of stiff, high modulus materials. For safe anchorage of a geomembrane system, also the loads resulting from damage in the geomembrane causing water to infiltrate behind it must be considered. This was found to be the most critical load. The behaviour of the two geomembrane systems was tested in the TUM canal, simulating damage in the geomembrane by cuts of different sizes made at different angles. Water pressure cells, water level indicators, force meters, bottom profilers, and propellers, were installed in the canal to measure the various parameters involved and necessary to verify behaviour of the system. The influence of the use of drainage valves to reduce the water pressure below the geomembrane was also investigated (Fig. 1). A significant reduction of forces was observed with the use of a valve to discharge under-pressure.

The results of the research by TUM gave a better understanding of the aspects involved when designing geomembrane system that must resist water in motion, and indicated that low modulus geomembranes require less anchorage lines than stiff geomembranes. The results perfectly match the design approach that had been adopted by the Carpi Group - specialised in design, supply and installation of waterproofing geomembrane systems in hydraulic structures - in previous projects in large canals, such as among others Mittlerer Isar-Strogenbauwerk canal in Germany and Laufnitz-dorf canal in Austria, both lined in 2000 with an exposed SIBELON® geocomposite. If the TUM guidelines indications will be followed within the community of engineers

Fig. 1. TUM research. Scheme and field execution of a cut in the PVC geomembrane perpendicular to the water flow and spanning the entire width between two fixation lines. No failure

involved in design of geomembrane sealing systems for canals, the comparative evaluation between different anchorage options will be correctly performed, decreasing the risk of inadequate applications and improving the performance of these systems.

2.2 Geomembrane Selection

Selection of the geomembrane is a critical choice deeply connected to the design of the fastening system, due to the above mentioned different ways of transferring loads to the anchors. Among the numerous synthetic materials available on the market, only those specifically designed for applications in large hydraulic structures and having documented long-term experience in such applications should be considered. ICOLD (2011) and USACE (1995) have investigated these aspects and given useful guidelines. Low modulus geomembranes heat-bonded during fabrication to an anti-puncture geotextile, such as SIBELON® geocomposites, have proven to be the most suitable in terms of performance, successful precedents, and constructability. Due to their flexibility, these geocomposites better conform and transfer to the substrate the loads, thus relieving the stresses on the fastening system. They have a superior performance in respect to puncture and burst, reliable and easy seamability, and high mechanical and weathering resistance, as proven by > 35 years installations in exposed position on demanding structures such as high dams, and canals with high water velocity. On the contrary, stiff geomembranes with low dimensional stability such as HDPE transfer excessive stresses to the fastening system (Fig. 2), are more difficult to place properly, are less performing in respect to puncture and burst by irregular substrates that are quite common in canals' rehabilitation, are prone to stress cracking, and seams are not totally reliable.

In the geocomposite, the geomembrane provides watertightness, and the geotextile provides anti-puncture protection further reducing the need for surface preparation, higher dimensional stability, improved mechanical characteristics, higher friction angle facilitating placement on the slopes, and some drainage capacity. Field experience has proven that a geocomposite is to be preferred to a geomembrane and a separate geotextile, not only because of shorter and easier installation: a separate geotextile does not assure good transfer of stresses to the substrate, there is a danger that excessive stresses

Fig. 2. Rupture of an exposed HDPE geomembrane where accidental damage to the HDPE resulted in water infiltrating under the geomembrane that transferred excessive stresses leading to failure of the fastening system

are transferred to the fastening system, and that in case of geomembrane damage the geotextile detaches and accumulates behind it.

The thickness of the geomembrane and the mass per unit area of the bonded geotextile are selected depending on required mechanical and environmental resistance, on the type of subgrade, and on whether the geomembrane will be left exposed or will be covered.

2.3   The Acting Loads and the Fastening Systems

The design of the fastening system must consider the loads imparted by water flowing inside the canal and the uplift caused by wind and by groundwater in the slopes when the canal is empty, and the uplift caused by flowing water infiltrating behind the geocomposite liner through accidental damage when the canal is full (Strobl and Schaefer 2005). The uplift by flowing water behind the geocomposite is generally the crucial design load. Environmental loads such as temperature, UV, water ingredients, and other similar site-specific agents influence the selection of the geocomposite nature and thickness.

Unless there are specific requirements, such as the need of machinery/animal passage or high risk of vandalism, the geocomposite is left exposed, to avoid reducing the canal section and to better exploit the low hydraulic roughness of the material. If the geocomposite is left exposed, the fastening system is designed site-specific, in function of the cross section of the canal, of the maximum water speed, of the maximum wind uplift, of under-pressure from the ground, of the bearing characteristics of the subgrade, and of the existing drainage system if any. For small irrigation canals the fastening system can be very simple and inexpensive, while more sophisticated design is needed for larger hydropower canals. A fastening system with high quality (stainless-steel is mandatory) profiles, suitably dimensioned and positioned, and in some cases designed to be capable of tensioning the liner and make it conform at best to the underlying structure, is generally adopted in canals of large cross section and/or with high water velocity, and in case increase in water flow is required. In canals with small cross section and/or low water velocity, anchorage can be less robust, and can be either linear or only at points.

Leaving the geomembrane exposed can offer the further advantages of eliminating/minimising vegetation growth, and of quick and easy maintenance, because any accidental damage can be more easily detected and repaired.

Fastening by a cover layer (ballasted geocomposite) can theoretically be made on the whole section of the canal, but in practice it is adopted in those areas where heavier service conditions are foreseen. The ballast layer can be concrete, generally cast over a protection geotextile placed on the PVC geocomposite to protect it against possible damage during casting of the concrete, or loose materials as further discussed.

2.4   Drainage

Drainage occurs in the gap provided behind the exposed geocomposite, and can be facilitated by installation of synthetic materials providing extra in-plane transmissivity when very high drainage flow is foreseen. Highly transmissive geonets are generally used for this purpose: they are installed under the waterproofing liner, only on the bottom of the canal, or over its whole cross section. Drained water discharge can be made into the existing drainage system of the canal, or inside the canal itself, by means of one-way valves activated by the difference in internal and external water pressure. An efficient drainage system has proven to be beneficial in reducing stresses on the anchorage system: a significant reduction of forces was observed with the use of drainage valves to reduce the water pressure underneath the geocomposite.

2.5   Sealing at Boundaries

One of the key issues of geomembrane sealing systems is how to fasten in a watertight way the geomembrane at boundaries, to avoid water infiltration. This aspect is of particular importance in case of hydraulic structures with water in motion.

At the beginning and end of the lined area, around the inlet or outlet zones, at spillways, at access stairs, and at any location where there is danger of water penetrating behind the geocomposite, a watertight seal must be made. If the seal is made on concrete or on masonry regularised with shotcrete, the seal is mechanical and watertightness is accomplished by compressing the liner over the substrate. On asphalt concrete, the seal is of the insert type and watertightness is accomplished by a watertight resin. In granular subgrades, trenches backfilled with impervious material are used.

3   Case Histories

The case histories that follow detail some typical applications, namely a canal where the geomembrane is totally exposed and anchored with a tensioning system, a canal where the geomembrane is totally exposed and anchored with a linear anchorage system without tensioning, and one canal where it is partially ballasted. All case histories discussed have used SIBELON® geocomposites.

3.1 Senhora Do Porto, Portugal 1994/1995

Senhora do Porto canal is a 2.5 km long hydropower canal with invert width from 4.1 to 5.1 m, and average height of 3.2 m. The initial design water flow, which was about 13 m³/s, had decreased due to due to loss of water through large fissures located at the springing of the rock masonry walls, and to the roughness of the lining, which had an estimated global Strickler-Manning coefficient = 55 m$^{1/3}$/s.

Objectives of the rehabilitation were to restore watertightness and to upgrade the canal so that it could transport the 20 m³/s flow that was necessary for maximum power output. The lining alternatives considered by the owner EDP – Energias de Portugal (at that time Electricidade de Portugal) were a 15 cm reinforced concrete liner, and a geomembrane system. The canal's walls were to be heightened of about 10% (30 cm). Hydraulic calculations showed that the geomembrane option allowed attaining an average increase in capacity about 50% higher than that of a new concrete liner. The geomembrane option was selected. The objective specified in the contract was to attain a global Manning Strickler coefficient = 85 m$^{1/3}$/s ± 5%, and to increase the capacity of the canal to the required 20 m³/s (Liberal and Ribeiro 1996). To attain this objective, design and installation aimed to guarantee that the liner would be smooth and plane over the substrate. A tensioning anchorage system and efficient drainage of uplifts were paramount.

The selected waterproofing liner was a geocomposite consisting of a 2.5 mm thick geomembrane heat-bonded during manufacturing to a 500 g/m² geotextile. Table 1 shows the main characteristics of the geocomposite. Adjacent geocomposite sheets were joined by watertight heat-seaming. The geocomposite is drained behind. Since high backpressure was possible in case of rapid drawdown of the canal, a high

Table 1. Senhora do Porto geocomposite

Property	Test method	Values & tolerances
Thickness (geomembrane only)	EN 1849/2	2.5 mm ± 5%
Specific gravity (geomembrane only)	EN ISO 1183/1 Method A	1.25 g/cm³ ± 4%
Test Performed on Geocomposite Sample Peak Value at Geotextile break* • Tensile strength • Elongation Peak Value at Geomembrane break* • Tensile strength • Elongation	EN ISO 527/4 (test speed 100 mm/minute)	≥ 45 kN/m ≥ 65 % ≥ 25 kN/m ≥ 250 %
Mass per unit area of geotextile	EN ISO 9864	500 g/m² ± 5%
Tear resistance (on nominal thickness of geomembrane)	EN ISO 34/1 (Specimen Speed 50 mm/min)	≥ 130 kN/m
Puncture resistance (PVC layer upwards) CBR	EN ISO 12236	≥ 7 kN

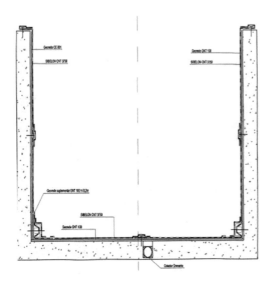

Fig. 3. Senhora do Porto: Tensioning profiles at bottom of walls, flat profiles at middle and top

transmissivity geonet was placed under the geocomposite on the walls and on the invert. The geonet conveys drained water to a fissured pipe installed in a longitudinal trench excavated at the invert (Fig. 4). Drainage discharge is made by transverse outlets positioned at 100 m spacing and crossing the invert and the downstream wall.

To achieve the increase in water flow required by contract, the fastening system had to be capable of keeping the geocomposite as taut as possible to the subgrade, with no folds. This was accomplished by a robust tensioning system developed and patented by Carpi. The anchorage system (Figs. 3 and 5) consists of one line of stainless-steel tensioning profiles installed at the base of each wall, of one line of stainless-steel flat batten strips at the invert, and one line of stainless-steel flat batten strips at the middle

Fig. 4. Senhora do Porto: From left to right, installation of drainage geonet, drainage collection pipe excavated at bottom, and the geocomposite installed over the geonet on the walls and part of the bottom

Fig. 5. Senhora do Porto: Detail of tensioning profile at bottom of walls, and view of the fastening system at bottom and middle and top of walls

and at the top of each wall. All fastening lines are waterproofed by a cover strip of geomembrane, the same material and thickness of the waterproofing geocomposite, watertight seamed upon the geocomposite along the perimeter of the fastening lines.

Total time required to perform civil works to heighten the walls, and to line 26,300 m², was 6 months, in the two dry seasons of 1994 and 1995.

Hydraulic measurements and calculations performed by the owner for acceptance confirmed that the global Manning-Strickler coefficient and the total capacity complied with specifications. Comparing data for the same level in the forebay reservoir, and after 10% heightening of the canal's walls, conclusions drawn by the owner (Liberal and Ribeiro 1996) were that the described drained geomembrane system was a clear technical success: maximum increase of capacity up to 90% was attained, and the system was very efficient from the point of view of waterproofing and of elimination of vegetation. Also from the financial point of view the system was very effective: besides being significantly less expensive than traditional systems, power production increase corresponded to a gain of more than 110,000 US$/year. The owner could pay back rehabilitation works in 4 years (Magalhães and Machado do Vale 2004).

3.2   Mittlerer Isar-Strogenbauwerk, Germany 2000

Mittlerer Isar is a 51 km long hydropower canal owned by E.ON (a large investor-owned energy service provider) which supplies water to four hydropower plants. In the Strogenbauwerk section, the canal is trapezoidal, 11 m wide at invert and 31 m wide at top, 6.5 m deep, and with a water velocity of 1.0 m/s. The canal was lined with concrete, which over the years deteriorated and resulted in leakage permeating the banks and, in the Strogenbauwerk section, affecting the safety of the road that crosses underneath the canal. Objective of the rehabilitation was to stop leakage, and assure the safety of the road. A monitoring system capable of locating the leaks if any was required by the owner.

The selected system was a drained geocomposite system. The liner is a geocomposite consisting of a 2.0 mm thick geomembrane heat-bonded during manufacturing to a 200 g/m² geotextile. Table 2 shows the main characteristics of the geocomposite. The liner was installed on the critical 900 m long Strogenbauwerk section.

Table 2. Mittlerer Isar-Strogenbauwerk geocomposite

Property	Test method	Values
Thickness (geomembrane only)	DIN 16726 par. 5.3.1 (DIN EN ISO 2286-3)	≥ 2.0 mm
Specific gravity (geomembrane only)	DIN 53479	≥ 1.25 g/cm^3
Tensile strength 1. ultimate load 2. strain at failure	DIN 16726 par.5.6.1. Table 1 - A - VII	≥ 22 kN/m ≥ 230 %
Mass per unit area of geotextile	DIN EN 29073 - 1 ISO 9073 - 1	≥ 200 g/m^2
Tear resistance	DIN 53363	≥ 200 N
Puncture resistance	DIN 16726 par. 5.1236	≥ 1100 mm

The 2.10 m wide geocomposite sheets were assembled in a prefabrication yard to construct 6 m wide panels, for quicker installation. Adjacent geocomposite panels were joined by watertight heat-seaming. The geocomposite was installed in exposed position, fastened by C-shaped stainless-steel profiles at the bottom and middle of the slopes, and by flat stainless-steel batten strips at the top of the slopes (Fig. 6). The C-shaped profiles are equipped with watertight fittings at geocomposite crossings, so geomembrane cover strips are not needed. Where the deterioration of the old concrete liner did not allow using the standard chemical anchors to fix the batten strips, deep grouted anchors were used. See Fig. 7 at left.

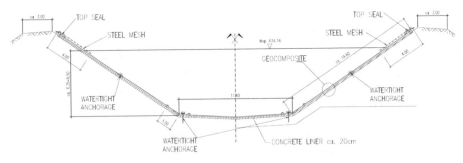

Fig. 6. Mittlerer Isar-Strogenbauwerk: C-shaped profiles at bottom and middle of slopes, flat profiles and steel mesh at top

Fig. 7. Mittlerer Isar-Strogenbauwerk: Drilling of deep anchors for the fastening system on bad concrete, geocomposite panel under installation, placement of the C-shaped profile at bottom of slope

The drainage layer is the geotextile associated to the geomembrane. Drainage water is collected at bottom by longitudinal strips of geonet and is discharged in the existing drainage system. To allow monitoring the performance of the liner, and in case there is a leak to locate the area of the leak, a system with optical fibres cables (OFC) was installed. The OFC system is activated by temperature differentials that occur when the presence of seeping water alters the "normal" temperature distribution. The OFC were placed at the bottom of the slopes on the existing concrete, under the geocomposite, and will allow locating the section where a leak should occur.

At the start and end of the lined stretch, a mechanical watertight transverse seal was installed (Fig. 8 at left): 80×8 mm stainless-steel flat profiles compress the geocomposite on the concrete regularised with epoxy resin, with rubber gaskets and stainless-steel splice plates to evenly distribute compression under the profiles and at abutting profiles. Before rehabilitation, the canal was equipped with steel grids every 100 m for safe exit of humans and animals. The owner required safe exits to be provided at least at the same spacing. The adopted solution installed over the geocomposite, from top of the slopes down under the minimum water level, a continuous steel mesh. This solution provided safe exit along the entire lined stretched of the canal. See Figs. 6 and 8, at middle.

Fig. 8. Mittlerer Isar-Strogenbauwerk: Mechanical seal at transverse boundary, placement of the steel mesh on the top part of the walls, to allow safe exit of humans and animals, and completed works

The installation of the drained geocomposite system was performed in 2000; the total installed quantity was of 34,560 m². No maintenance has been required up to present year 2018; the imperviousness of the geocomposite is unaltered.

3.3 Tekapo, New Zealand 2013/2014

The Tekapo Canal transfers water 25.3 km from the tailrace of the Tekapo A Power Station, on the edge of Lake Tekapo, to the head pond of the Tekapo B Power Station, which discharges into Lake Pukaki in the South Island of New Zealand. The canal has a capacity of 130 m³/s and a water velocity of approximately 1.2 m/s. The water depth of the canal varies from 5.3 to 6.4 m, the top water width varies between 33 m and 37 m. Leakage and evidence of erosion of the original earth lining of the canal (construction was completed in 1977) led to the need for repairs. Throughout 2011 various possible remedial solutions were investigated by the Owner, Genesis Energy. Primary objectives were a 50 years design life, a well proven track record, the capability to optimise the time for construction minimising outage, and a high degree of confidence in technical performance of the new liner, because the canal is in a highly seismic zone. In 2012 Carpi's geomembrane system was selected based on its superior technical performance and on the capability to attain also the other objectives.

The Tekapo Canal Remediation Works included lining three selected sections of the canal with the geomembrane system, to improve the long-term reliability and performance of the canal, structural strengthening of the State Highway bridge crossing the canal upgrading its seismic resilience, and the reconstruction of the entire embankment of a short stretch of the canal to allow the replacement of an underlying drainage culvert.

The cross-section of the remediated stretches is represented in Fig. 9. The waterproofing liner is the same adopted at Senhora do Porto (see Table 1). Adjacent geocomposite sheets were joined by watertight heat-seaming (Fig. 12). The geocomposite is anchored by ballast on the invert and on the upper part of the slopes. A layer of cobbles, sourced locally, was placed to form the invert ballast, and a zone of cobbles

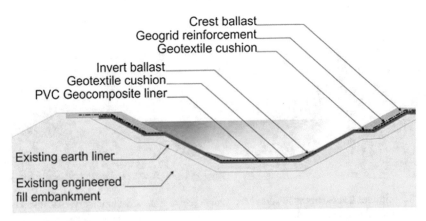

Fig. 9. Tekapo: Cross section of the new geocomposite liner

with a geogrid reinforcing layer on the upper part of the slopes to form the crest ballast on each side of the canal. A 2,000 g/m^2 non-woven needle punched geotextile was placed over the geocomposite in the invert of the canal to protect it from damage during the placement of the cobble invert ballast. The same type of geotextile, with a 1,000 g/m^2 mass per unit area, was placed over the geocomposite on the upper part of the slopes to protect it from the placement of the crest ballast system (see Figs. 10 and 11). The transverse watertight seal at the start and end of the lined stretches was made by inserting the geocomposite in an excavated trench then filled with plastic low-strength concrete (Fig. 13 at right).

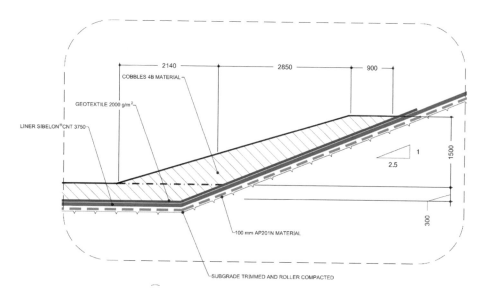

Fig. 10. Tekapo: Detail at invert

With the need to shut down both Tekapo A and B Power Stations to dewater the canal to carry out the liner repair works, and the consequent loss of electricity generation, there was a requirement to carry out the liner works in the shortest possible length of time. To achieve this, it was decided to adopt an approach of early contractor involvement in the design process. This commenced with an integrated owner/designer/contractor design workshop nearly one year before the start of construction. Prior to final design, full scale trials were carried out to assess some critical construction techniques, prove the design and demonstrate QA/QC procedures: bottom ballast, liner system placement and slopes ballast, and cofferdam construction and removal. All trials were successful. The sequence of works consisted in construction of cofferdams across the canal, dewatering and fish removal, subgrade preparation to achieve a stable subgrade without excessive protrusions, installation of the geomembrane system, installation of ballast on the invert and at the top of the slopes, re-watering of the canal sections, removal of cofferdams and return to service. Sixteen people were dedicated to Quality Management.

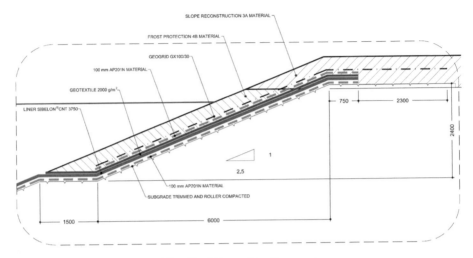

Fig. 11. Tekapo: Detail at crest

Fig. 12. Tekapo: Placement and seaming of the new geocomposite liner

Fig. 13. Tekapo: Placement of ballast on invert, transverse perimeter sealing

The works were carried out in two separate stages, the first one in 2013 and the second one in 2014. It was planned to carry out Stage 1 of the remediation works during a 14-week electricity outage. Following the long period of careful planning and the pre-construction site trials, these works were successfully carried out in 12 weeks, on 5.73 meters of canal length and 270,000 m^2 of liner. In the second season, works were also completed 2 weeks ahead of time, in 8 weeks, on 1.425 m of canal length and 80,000 m^2 of liner (Fig. 14 at right).

Fig. 14. Tekapo: Placement of ballast on upper side of slope, works completed

4 Lining Canals Without Stopping the Water Flow

Many canals cannot be dewatered, or dewatering entails excessive inconveniences and costs. The available underwater technologies developed and installed by Carpi in high dams and large reservoirs cannot be applied in most canals, because divers can safely operate only in still water, or in water flowing at speed <0.5 m/s. Developing a geomembrane system that could be installed in flowing water has been a milestone in the field of waterproofing techniques for hydraulic structures. The system has been developed in the last few years and has already been installed in two pilot projects. The system adopts a mattress, whose trademark is SIBELONMAT®, which consists of two watertight geomembranes connected to form a mattress, with a connection system allowing free distribution of a filling material, such as inexpensive cement grout, which provides the ballast anchorage for the mattress. The bottom geomembrane is the geocomposite already discussed, which provides the waterproofing function, the upper geomembrane is a geotechnical PVC-coated material whose function it to confine the grout inside the mattress, avoiding that dispersion of cement may pollute the water of the canal. The solution is a no-environmental and no-operational impact solution. The upper geomembrane also contributes to improve the hydraulic efficiency of the mattress.

The mattresses are prefabricated in panels of 10 m width and predefined length to minimise junctions and facilitate placement. During prefabrication, watertight heavy-duty zippers are sealed to the panels, for underwater joining of adjacent panels (Fig. 15). The panels are equipped with devices for watertight insertion of the grouting pipes.

Fig. 15. SIBELONMAT® mattresses under preparation, watertight zipper, SIBELONMAT® mattress filled with cement grout

Other types of mattresses are available in the market, consisting of fabrics filled with micro cement, prefabricated in panels and joined in the dry with zippers. The watertightness is provided by the thickness and cement content of the filling. The disadvantages of such mattresses are that the leaching of the cement during the filling may contaminate the water, and that the watertightness of the zippers is not proven. If filling of the panels is not carried on continuously, cold joints will form, the inevitable shrinkage of the filling will create cracks, the non-watertight connection between adjacent mattresses will allow infiltration of water. In summary, fabric mattresses filled with cement are suitable for erosion control but are not able to provide long-term watertightness.

The mattress here discussed differs from fabric mattresses because the watertightness is provided by the geomembrane and the watertight zippers, and the cement grout provides only the ballast. Therefore, less expensive grout can be used and the behaviour of the grout will not affect the watertight performance of the mattress. Extensive full-scale testing has been made in the dry, underwater with still water, and underwater with water in motion, to develop a dependable solution. The system was then installed in a section of a canal in Italy and in a section of the Ismailia canal in Egypt.

Ismailia canal was constructed in 1862 between the cities of Zagazig and Ismailia to carry fresh water to Suez Canal cities. The canal, 128 km long, is for much of its length constructed through sandy strata. The canal has successively been widened and the current width varies from 30 to 50 m. Expansion and canal widening has resulted in water-logging and salinization, especially along areas adjacent to the widened stretches of the canal. As the Nile River no longer carries and deposits substantial quantities of sediment after the construction of the High Aswan Dam, progressive widening of the canal removes layers of less permeable silt without subsequent replacement, thus resulting in increased seepage. Other problems range from deteriorating irrigation infrastructure such as irrigation offtakes, regulators, canal obstacles such as bridges, etc. It is against this background that the Government of Egypt is considering rehabilitation of the Ismailia canal, potentially with the watertight mattress system. To assess and validate the technology, a 25 m long section of the canal was lined with three panels, of which two are 10 m wide and the third is 5 m wide. Each mattress was

75 m long and spanned the canal cross section from the crest of one slope to the crest of the opposite slope. Installation was performed in flowing water. The bottom of the mattress was grouted first, up to the bottom part of the slopes. When the bottom part had cured enough to avoid that the grout of the slopes would deposit also on the bottom, the slopes were grouted. See Fig. 16.

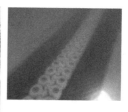

Fig. 16. Ismailia: From left to right, installation of the SIBELONMAT® mattress, the filled mattress, and the watertight heavy-duty zipper

In both installations there was no leaching of grout from the mattresses, which is mandatory from an environmental standpoint. In both cases the injection of the grout was carried out in the dry but methodology is available for underwater grouting.

The benefits of the system are that it can be installed underwater with no disruption of operation, in rather short installation times, say 1500 m^2 surface lined by one crew in a day. The synthetic materials that form the mattress are extremely flexible and have very high mechanical properties. They can adapt to rather rough supporting surfaces, without need of excessive surface preparation. The system is wholly watertight: watertightness is provided by the bottom geocomposite, formulated to provide an extensive functional life (it is proven that functional life of a 3 mm thick geomembrane, exposed, is >100 year, see Giroud 2013), and by the watertight zippers connecting adjacent panels. The grout has no waterproofing function, but only a ballasting function. Cracking in the hardened cement grout is no issue for the efficiency of the system, therefore low strength grouts may be used. Grout is not dispersed in the water during filling of the mattress, so preserving the environment. The system, conceived for canals, can be used for lining embankment dams and for blanketing.

5   Conclusions

Lining systems with geocomposites provide durable continuous watertightness and can substitute traditional rigid liners providing technical and financial benefits. They can bridge joints and large fissures and resist opening of new fissures in case of settlements in the natural slopes. The hydraulic roughness, much lower than that of traditional materials, is maintained over time and allows increasing the water flow. Installation times and costs are lower than those of other rehabilitation systems, practically no maintenance is required. The SIBELONMAT® system will provide further advantages in canals' rehabilitation.

References

Christensen, J.C., et al.: A conceptual design for underwater installation of geomembrane systems on concrete hydraulic structures. Technical Report REMR-CS-50 (1995)

Durand, B., Tremblay, S.: Study of waterproofing revetments for the upstream face of concrete dams. Final Report IREQ 95–326 (1995)

Giroud, J.: Functional Service Life of SIBELON® Geomembrane for Panama Canal. Memorandum (2013)

ICOLD, The International Commission on Large Dams: Bulletin 135: Geomembrane Sealing Systems for Dams—Design Principles and Review of Experience (2011)

Liberal, O., Ribeiro, V.: Remodelação do canal da Senhora do Porto. In: Proceedings, Utilização de geomembranas em hobras hidraulicas e subterrâneas, First Portuguese National Conference—Laboratorio Nacional de Engenharia Civil, Lisbon, Portugal (1996)

Magalhães, P.M., Machado do Vale, J.L.: Uprating and refurbishment of Senhora do Porto power plant's headrace canal. In: Hydro 2004—A New Era for Hydropower, Porto, Portugal (2004)

Plusquellec, H.: Application of geosynthetics in irrigation and drainage projects. ICID-CIID, International Commission on Irrigation and Drainage, New Delhi (2004)

Schaefer, P.: Basic research on rehabilitation of aged free flow canals with geomembranes. Berichte des Lehrstuhls und der Versuchsanstalt für Wasserbau und Wasserwirtschaft der Technischen Universität München, Nr. 107 (2006)

Strobl, T., Schaefer, P.: Exposed thermoplastic geomembranes for sealing of water conveyance canals, guidelines for design, supply and installation. Berichte des Lehrstuhls und der Versuchsanstalt für Wasserbau und Wasserwirtschaft der Technischen Universität München, Nr. 105 (2005)

Rehabilitation of Dams with Watertight Geomembranes, in the Dry and Underwater

Alberto Scuero[1](✉) and Gabriella Vaschetti[2](✉)

[1] Carpi Group, Balerna, Switzerland
alberto.scuero@carpitech.com
[2] Carpi Tech, Balerna, Switzerland
gabriella.vaschetti@carpitech.com

Abstract. Geomembranes are watertight synthetic materials, factory-manufactured under computer-controlled procedures that guarantee constant properties and quality, and supplied as flexible sheets a few millimetres thick. Geomembranes are used to waterproof all types of hydraulic structures, for rehabilitation and for new construction, in the dry and underwater. In dams, geomembranes were first used at the end of the 1950ies. This paper focuses on rehabilitation, discussing the available options for concrete and embankment dams, and for RCC dams. The state-of-the-art systems place the geomembrane liner at the upstream face of the dam, generally in exposed position, and with a drainage system behind, to allow discharging possible backpressure, and monitoring the performance of the system. This flexible technology can be adopted to line the entire upstream face of the dam, or only the area/s causing most seepage, or only cracks/failing joints. Projects include all types of dams and subgrades (concrete, bituminous concrete, granular subgrades or RCC). Geomembrane systems do not require heavy equipment and site organization. Large use other types of geosynthetics is made to perform functions that with traditional materials would require complicated and time-consuming operations: thick unwoven geotextiles to provide anti-puncture protection avoiding surface preparation, geogrids or high-performance technical woven geotextiles to provide support over cavities in the subgrade, geonets to enhance drainage capacity. The face anchorage and peripheral anchorage are calculated to resist acting loads such as uplift by wind and waves. Face anchorage on solid subgrades is generally linear, with a tensioning system, while on granular subgrade it is made at points, with deep anchors. Ballast anchorage is rarely used. The paper presents case histories of repairs in the dry and underwater.

1 Introduction

Dams have since centuries been built to retain water for times of scarcity, and to control water when it is in excess, so as to avoid damages to people, land, property and infrastructures, and in modern times to produce electricity. Water is an essential but not unlimited resource for life on our planet. The dramatic increase in population, and climate changes, will make water an increasingly precious albeit increasingly scarce resource in the decades to come. For this reason, dams and their appurtenant structures

© Springer Nature Switzerland AG 2019
M. Meguid et al. (Eds.): GeoMEast 2018, SUCI, pp. 77–96, 2019.
https://doi.org/10.1007/978-3-030-01944-0_7

must be adequately designed, constructed, and maintained, to ensure their safe operation.

In many countries many dams are very old; ageing almost always results in decreased watertightness, which may become a hazard for the safety of the dam. In concrete and masonry dams, uncontrolled seepage may result in deterioration of the cementitious material and consequent reduction of the material's strength. If the infiltrating water is not properly drained and discharged, uplift pressures may jeopardize the overall dam stability. In embankment dams, uncontrolled presence of water inside the dam may trigger erosion, piping, and embankment instability, which may ultimately lead to failure.

Over the centuries, several techniques have been used to restore watertightness of dams; most used materials for local repairs have been mortar, shotcrete, concrete, resins, while re-facing has been done with concrete, shotcrete, or asphalt concrete. Local repair techniques, which in the short term may be efficient in reducing water seepage, are not effective in the long term, and the need for frequent repeated repairs may at the end be very costly, especially if repairs involve lowering the water level. Re-facing requires dewatering of the reservoir, with long outage, high impact on water supply and power production, and consequently also high financial impact. After World War II, dam engineers started exploring and experimenting new systems based on the use of synthetic, factory-produced and practically watertight materials that had become available on the market. Such materials, known as geomembranes, are manufactured in a few millimetres thick films, and being produced under stringent computer-controlled quality control procedures have constant characteristics, and are always available in the needed quality and quantity.

Geomembranes started to be used as water barriers in dams since 1959. Their physical and mechanical properties can be designed and controlled all along the process of production in the factory; the final quality of the product is not influenced by weather conditions at the time of installation, which on the contrary applies to other repair technologies such as concrete, asphalt concrete, or painted or sprayed products. Furthermore, different from traditional repair methods, installation of geomembranes has minimum environmental impact, as it does not require a production plant at site nor heavy equipment for transport and handling. Additionally, installation of geomembranes is quicker than other rehabilitation technologies, it can be made in stages, or underwater, so with minimum or no impact on operation of the dam. Geomembranes do not require routine maintenance and, in case of accidental damage, they may be repaired with some patch work, by the owner's personnel duly trained.

Geomembranes are on one side an established technology with proven long-term durability; on the other side the technology is continuously evolving to meet specific needs of dams' owners. They can be used for new construction of Roller Compacted Concrete (RCC) dams, of embankment dams and of cofferdams, and for rehabilitation of old leaking dams. In rehabilitation, which is the focus of this paper, they have been used on all types of dams: concrete gravity, buttress, arch, and multiple arches dams, masonry dams, rockfill dams, earthfill dams, and RCC dams. The geomembrane system is placed at the upstream face, possibly in combination with an efficient face drainage system; it can cover the whole face of the dam, or just parts of it, or only failing joints or cracks. The following chapters address the geomembrane solutions adopted on different types of dams, for total rehabilitation and for local repairs.

2 Design Guidelines and Issues

The most authoritative international body dealing with dams, ICOLD, the International Commission on Large Dams, has since the end of the 1970ies been addressing the use of geomembrane in dams. Dedicated bulletins have been issued to provide information and guidelines: Bulletin 38, "Use of thin membranes in fill dams", published in 1981, Bulletin 78, "Watertight geomembranes for dams – State of the art", published in 1991, and Bulletin 135, "Geomembrane Sealing Systems for dams – Design principles and review of experience", published in 2010. The same authors have presented such bulletins at GeoMEast (Scuero and Vaschetti 2017).

Bulletin 135 (ICOLD 2011) gives extensive information on characteristics, applications and behaviour of all types of geomembranes, and guidelines addressing all aspects of design and installation; guidance to technical contents of contracts are included in the bulletin. Related bibliography and database on dams with geomembrane systems have been prepared concurrent with the Bulletin.

2.1 Geomembrane Selection

Several types of geomembranes are available, belonging to two main families: polymeric geomembranes and bituminous geomembranes. Since bituminous geomembranes are used only on embankment dams, and their application is limited and less and less frequent, they are not discussed in this paper.

Selection of the type and thickness of a polymeric geomembrane for dam rehabilitation must be based on site's specific requirements, on mechanical and physical properties, and on documented successful long-term experience in similar applications. In addition to the indications contained in ICOLD Bulletin 135, useful information was provided by USACE, the US Army Corps of Engineers, in a 2-years research programme that in its first phase investigated the geomembranes available on the market and performed large-scale tri-axal tests on 21 types of geomembranes and of composite geomembranes (geocomposites formed by a watertight geomembrane heat-bonded at manufacturing to an anti-puncture geotextile). Tests for resistance to puncture under repeated loading and unloading cycles in pressure vessels under 1 MPa water head (Fig. 1) and for resistance to burst (Fig. 2) indicated that low modulus PolyVinylChloride (PVC) geomembranes heat-bonded during fabrication to an anti-puncture geotextile, such as SIBELON® geocomposites, have superior performance than high modulus geomembranes such as High-Density Polyethylene (HDPE). Such geocomposites scored 55 out of 55, HDPE scored 7 out of 55. Assessment of other factors such as tear resistance, seamability, dimensional stability, overall constructability, precedents, durability, repairability, confirmed these geocomposites as the most performing, and HDPE geomembranes as the least performing, materials (Christensen et al. 1995). ICOLD statistics and field experience support such findings.

Fig. 1. Full-scale testing for puncture resistance by the US Army Corps of Engineers. Geocomposite Sibelon® CNT 3750 (2.5 mm thick geomembrane + 500 g/m² geotextile, top right) conformed to substrate resisting repeated loading-unloading cycles at 1 MPa (100 m). HDPE geomembrane of the same thickness (bottom right) did not conform and ruptured at 0.35 MPa at first loading

Fig. 2. Full-scale testing for burst resistance by the US Army Corps of Engineers. Geocomposite Sibelon® CNT 4600 (3 mm geomembrane + 700 g/m² geotextile, left) resisted 160 kPa, with a 271% surface elongation. HDPE geomembrane (right) failed at lower elongation, along a line indicating a zone of lower strength, and a preferential path for propagation of tear

Flexibility and elongation capability are paramount for the selection of a polymeric geomembrane: a flexible material will better adapt to the subgrade distributing loads and minimising the risk of puncture at protrusions and burst at cavities, good elongation will allow bridging opening of cracks/joints and adapting to settlements/differential movements. The flexibility and elongation capability of a geomembrane are illustrated by its tension-elongation curve, which allows assessing behaviour and hence the maximum tensile strength and corresponding maximum elongation (strain). In most typical situations in dams, maximum elongations induced in geomembranes in the field are in the order of 50–60%. Geomembranes that exhibit a peak or plateau at elongations lower than 50–60% are prone to premature failure in case of local thickness reduction caused by a scratch (Giroud 1984). Since it is well known that during installation geomembranes are scratched and local thickness reduction does occur, geomembranes should have a tension-elongation diagram increasing monotonically up to 50–60% elongation, while a peak or a plateau in the tension-elongation curve between zero and 50–60% elongation is unacceptable. The following paragraphs discuss the curves of HDPE and PVC geomembranes respectively.

The tension-elongation curve of a 2.5 mm HDPE geomembrane must be restricted to the range of admissible strains in the field, i.e. for an allowable elongation up to the yield point in the HDPE, which occurs at about 12% elongation (peak elongation), beyond which the behaviour becomes plastic. The yield point is in fact a point of instability because once it is reached the geomembrane thins down locally and elongates like gum, presenting a plastic elongation under essentially constant tension up to the elongation at break (at about 700% elongation). Beyond the peak, the HDPE geomembrane ceases to function from a mechanical standpoint (Giroud 1984). Therefore, HDPE geomembranes should be used only where the geomembrane elongation is well below the yield elongation, and with a substantial factor of safety. International literature (e.g. Peggs et al. 2005) indicates that to be on the safe side the allowable elongation of HDPE geomembranes should not exceed 3 to 6%, and that for elongations greater than 3% the "creep" phenomenon (irreversible deformation after the stress has ceased) is important and cannot be neglected.

The typical tension-elongation curve of a SIBELON® geocomposite with a 2.5 mm thick geomembrane and a 500 g/m^2 backing geotextile shows a monotonically increasing diagram with two peaks. The first peak (at about 65% elongation) corresponds to the break of the backing geotextile, and the second peak (at >250% elongation) corresponds to the break of the geomembrane. Beyond the first peak, the material presents the characteristic behaviour of the geomembrane until failure.

Comparing the tension-elongation curves (Fig. 3) gives a good insight of the suitability the above-mentioned materials for application in dams. Clearly, HDPE geomembranes are not adequate for use in dams. Their high (>700%) elongation at failure must not be taken as indicator, because it is the yield point that dictates the behaviour: elongations >3–6% cannot be accepted. In contrast, the curve of the PVC geocomposite, increasing monotonically from zero up to 50–60%, shows that it can be used in situations where it will experience considerable deformations, well above those that can be reasonably expected in the field: it can accommodate high deformations of the subgrade, differential movements, opening of joints, and repeated stresses exerted by waves and wind, with no risk of breaking.

In the geocomposite, the flexible elastic geomembrane provides the water barrier (permeability < 10^{-6} m^3/m^2/day as per EN 14150 standard), and the geotextile has several functions: provide anti-puncture protection, enhance the mechanical properties of the liner, improve dimensional stability, provide higher friction angle facilitating placement on inclined facings, and provide some drainage capacity. Flexibility is an important characteristic that allows the geomembrane conforming to the subgrade, distributing loads and adapting to complex geometries. This results in high resistance to damage during installation, in high resistance to puncture and burst, in capability of bridging cracks and fissures, in capability of accommodating differential displacements, in reduced formation of wrinkles.

The thickness of the geomembrane and the mass per unit area of the bonded geotextile are selected depending on required mechanical and environmental resistance (UV radiation and extreme temperatures being the heaviest loads), on the type of subgrade and water head, and on whether the geomembrane will be left exposed or will

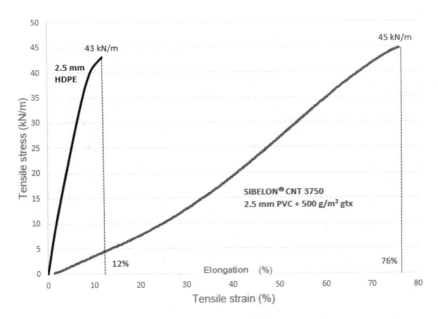

Fig. 3. Tension-elongation curve of a 2.5 mm thick HDPE geomembrane compared to tension strain curve of a SIBELON® CNT 3750 geocomposite (2.5 mm thick geomembrane + 500 g/m² geotextile heat-bonded at fabrication)

be covered. Typically, in dams an exposed geomembrane has a thickness not inferior to 2.5 mm. Table 1 shows some typical values that should be required for a geocomposite to be adopted in such installations.

Table 1. Characteristics

Property	Test method	Values & tolerances
Thickness (geomembrane only)	EN 1849/2	2.5 mm ± 5%
Specific gravity (geomembrane only)	EN ISO 1183/1 Method A	1.25 g/cm³ ± 4%
Test Performed on Geocomposite Sample Peak Value at Geotextile break* • Tensile strength • Elongation Peak Value at Geomembrane break* • Tensile strength • Elongation	EN ISO 527/4 (test speed 100 mm/minute)	≥ 45 kN/m ≥ 65 % ≥ 25 kN/m ≥ 250 %
Mass per unit area of geotextile	EN ISO 9864	500 g/m² ± 5%
Tear resistance (on nominal thickness of geomembrane)	EN ISO 34/1 (Specimen Speed 50 mm/min)	≥ 130 kN/m
Puncture resistance (PVC layer upwards) CBR	EN ISO 12236	≥ 7 kN

All case histories discussed in this paper have used SIBELON® geocomposites, whose geomembrane is stabilised against degradation from UV radiation so that it can be left exposed also in aggressive environments. Installations made at high altitudes in the Italian Alps more than 30 years ago testify that those geomembranes are still fully functional and their watertightness is unchanged. It is of essence however that geomembranes selected for dams' projects have adequate formulation and thickness to grant high durable performance, have successful precedents, and proven durability in the field.

2.2   Face Anchorage and Drainage

In dams' rehabilitation, the geocomposite is typically left exposed, so a face anchorage is needed to keep it in place against uplift caused by wind and waves. Several face anchorage systems are available and discussed in the case histories that follow; their selection is based on the type of subgrade (rigid with good pull-out resistance, or with low pull-out resistance, or loose material), on the geometry of the dam, on the need to tension the geocomposite, so as to remove wrinkles and slacks, which may become locations of stress concentrations and local premature aging of the geomembrane.

The state-of-the art system is based on the concept of a geocomposite having behind a full-face drainage system. Drainage occurs in the gap provided between the dam face and the exposed geocomposite. In some cases, a synthetic drainage layer, such as a high-transmissivity geonet or a draining geocomposite, is installed under the waterproofing liner. If a face tensioning system is used, the vertical lines that constitute the face anchorage form parallel vertical drains at the dam face, contributing to conveying by gravity the drained water to a longitudinal bottom collector. Water that should infiltrate beneath the geocomposite in case of accidental damage, or water already permeating the dam and migrating to the upstream face under the action of solar radiation, is captured in the drainage system, by gravity flows into the bottom collector, and from there to one or more transverse drainage pipes that discharge it in the gallery if existing, or downstream. Upstream discharge by drainage valves activated by pressure differential is also possible.

If the water pressure underneath the geocomposite is not balanced by the water pressure above the geocomposite, as in case of fluctuation in the water level of the reservoir or rapid drawdown, the geocomposite could bulge and be stressed. The face drainage system ensures that no stress is transferred on the geocomposite, avoids that water in pressure from the reservoir enters into the dam body, allows monitoring the behaviour of the geocomposite system, and can gradually reduce the water content in the dam. Reducing the water content in the dam can be very beneficial to for the safety of the dam, especially in case of possible or ongoing Alkali-Aggregate Reaction (AAR).

2.3   Peripheral Sealing

Peripheral sealing maintains the geocomposite in place at peripheries, such as along the bottom boundary, spillway, inlets, outlets, and prevents water infiltration underneath the geocomposite.

In dams, the peripheral seal is in most cases made on concrete or on masonry regularised with shotcrete, and is mechanical: watertightness is accomplished by compressing the liner over the substrate, regularised with appropriate resin, and using rubber gaskets and splice plates. On less rigid subgrades or on granular subgrades, slots/trenches filled with impervious material are used.

3 Full-Face Dry Applications

The case histories that follow describe the possible systems in typical applications: on a concrete dam, on a masonry dam, on an asphalt concrete facing fill dam, and on an earthfill dam. The applications to Roller Compacted Concrete (RCC) dams and to Concrete Face Rockfill Dams (CFRDs) are addressed in the chapters on waterproofing of crucial areas and on underwater waterproofing respectively.

3.1 Concrete Dams: Chambon, France

Concrete dams have a hard subgrade that allows using the tensioning assembly cited in ICOLD Bulletin 135, anchoring it to the concrete with stainless steel anchor rods embedded in chemical phials. The assembly consists of two stainless steel profiles: the first profile, U-shaped, is anchored to the concrete, the geocomposite sheets are placed overlapping on it, and are then covered by the second profile, Omega-shaped, which is fastened to the first one achieving fixation and tensioning of the geocomposite, as outlined in the conceptual scheme of Fig. 4, an excerpt of ICOLD Bulletin 135. The profiles, besides providing a pre-tensioning effect avoiding formation of slack areas and folds, participate to the drainage system as discussed in the previous chapter.

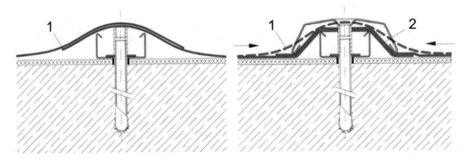

Fig. 4. Tensioning system: On the left, adjacent geocomposite sheets (continuous red line, 1) overlap on the U profile secured to the dam. On the right, the Omega profile placed on the geocomposite sheets compresses and forces them to change from the previous loose configuration (dotted red line, 1) into a new pre-tensioned configuration (continuous red line, 2). Excerpt of ICOLD Bulletin 135, with geocomposite marked in red for easier understanding

An outstanding example of this type of rehabilitation is Chambon dam in France, 137 m high, a concrete gravity dam made of cyclopean concrete, located in the French

Alps, at 1,042 m a.s.l. Owner of the dam, which used for hydropower, is EDF, the French National Power Board. The dam, constructed in the period 1929–1935, started to show evidence of AAR in 1958: cracking, opening of the construction joints at the upstream and downstream face, deformation of the curved section of the dam in the upstream direction. Rehabilitation works started in 1990 with the aim to grant stability and safety of the dam, in view of its future decommissioning and of building downstream a new larger dam, whose construction was estimated to start around year 2000. Rehabilitation works included grouting of cracks, sealing of the openings in the old spillway and construction of a new underground spillway, and a series of slot cutting to be performed over the years for tension relief. To prevent building up of pore pressures in the dam, in case of opening of the construction joints under earthquake, and to restore imperviousness after slot cutting, an exposed geomembrane system was installed in the upper 40 m of the dam, above the concrete wall at heel. A pilot installation was made in 1991 at the left abutment, and following 2 years of monitoring and good performance, in 1993 the exposed geomembrane system was installed on the entire upstream face.

The waterproofing geocomposite is formed by a 2.5 mm thick geomembrane heat-bonded to a 500 g/m^2 geotextile, and fastened to the dam face with the tensioning profiles described. The drainage system consists of a drainage geonet placed under the geocomposite, of the vertical drains constructed by the U-profiles, of a bottom collector in form of a perforated box drain (Fig. 5 at left), and of transverse pipes that discharge into the gallery. Separate drainage compartments were created to better monitor the performance of the system. The tensioning profiles, placed in parallel vertical lines at 1.8 m spacing, are waterproofed with a cover strip of geomembrane of the same material forming the geocomposite Fig. 5 at middle); the peripheral watertight seals are made with an 80 × 8 mm stainless-steel flat profile. The typical installation sequence of such seal is: placement of anchor rods, regularisation of concrete with an epoxy resin eliminating any residual roughness, puncturing of the geocomposite deprived of the geotextile on the rods, placement of an 80 × 3 mm EPDM rubber gasket and stainless-steel splice plates at abutting profiles to ensure the even compression necessary to achieve a watertight seal, placement of the profile, tightening of the bolts, control of the torque.

Fig. 5. Chambon: At left, the drainage geonet (the black material), the U-profiles forming parallel vertical drains at the upstream face, and the bottom box drain collecting and conveying drainage water to transverse drainage discharge pipes. At middle, the vertical tensioning profiles waterproofed with geomembrane cover strips, and the watertight bottom seal. At right, the geocomposite waterproofing the area where the slot has been cut to relieve tension due to AAR

When EDF needed to perform slot cuts for tension relief, the geocomposite was cut in the area of the slot, cutting was performed, and watertightness was quickly and effectively restored by installing a strip of geocomposite on the area of the cut (Fig. 5 at right).

For twenty years, the exposed geomembrane system performed as expected, protecting the dam from water infiltration. In 2013, to further extend the life of the dam, EDF decided to perform structural reinforcement of the dam body with pre-stressed tendons crossing from upstream to downstream, and with a net of carbon composite bands on the upstream face. New slot cuts were also planned. The structural rehabilitation works required dismantling the geomembrane system installed in 1993. For future protection of the dam from water infiltration, EDF required the same geomembrane system to be restored. Chambon is the first, and so far the only, example of a geomembrane system that, after 20 years' excellent performance, was removed to allow performing structural works at the dam, and then re-installed to protect the dam again.

The good conditions of the materials allowed re-using 40% of the drainage geonet and 55% of the fixations. Challenges were to synchronize the works of numerous contractors, to execute the works while the dam was operating, and without closing the crest road that connects France to Italy. Furthermore, being the dam close to a national park, the works had to have the minimum possible environmental impact. Finally, the works had to be performed in the shortest possible time and in harsh climatic conditions (Fig. 6 at left).

Fig. 6. At left, the upstream face of Chambon dam during the concurrent structural rehabilitation and waterproofing works, in winter. At right, the dam two years after completion of the 2013 waterproofing works

Chambon has been given other decades of service life, and for the moment construction of the new dam downstream has been postponed.

3.2   Masonry Dams: Kadamparai, India

According to ICOLD (2000), about 20% of the world's dams are masonry dams. These dams are particularly vulnerable because masonry has higher permeability than concrete, due to the numerous joints between masonry blocks, and lower tensile strength, both accentuated by ageing processes. ICOLD reports masonry dams as accounting for 80% of the failures of gravity dams, due to poor quality of water-tightness and water percolation through the dam body, which can dissolve the cementitious component of the mortar, with a decrease of watertightness. Local repairs are a short-term measure, while geomembrane systems are particularly effective in solving the problem of leaking joints between masonry blocks, and for this reason have extensively been adopted on this type of dam.

Kadamparai dam, 67 m high, is a significant example. The dam, owned by Tamil Nadu Electricity Board (TNEB) and used for hydropower, is a composite structure consisting of a central stone masonry gravity dam with earth embankment abutments. Completed in 1983, after first impoundment the dam exhibited 1,120 l/min seepage, which remained within the allowable limits for the first two years of operation, but starting from 1995 gradually increased. Repeated local repairs like racking and pack-ing, pointing, grouting, including chemical grouting, shotcreting etc. showed to be palliatives; leakage persisted and increased over the years, reaching the unacceptable rate of 38,000 l/min. Since the conventional methods already adopted had failed at Kadamparai, TNEB decided to seek a long-term solution to the problem. Various entities were consulted, and the final decision was to adopt an upstream geomembrane system. An international tender was issued in 2003, awarded in July 2004, and installation was carried out in 2005. To reduce the outage time, it was decided to install the geomembrane system only on the central structure, and exclude from the water-proofing works the gravity section covered by the sloping earth embankment sections.

The waterproofing geocomposite is a 2.5 mm thick geomembrane heat-bonded to a 500 g/m^2 geotextile. Face anchorage is made with parallel lines of the tensioning system escribed, placed at 5.7 m spacing. Perimeter sealing is watertight against water in pressure at submersible peripheries (bottom and abutments, spillway, and trash-racks at the scour vent tower), and watertight against rain and waves above maximum water level (dam crest, including scour vent tower). Peripheral watertight seals are the same described for Chambon. At bottom the seal is double: a primary seal is placed almost at the bottom of the upstream face, and a secondary seal is placed underneath it, to reduce the water head on the primary seal and create a separate drainage system to intercept water that should be seeping from foundations. Perimeter seals watertight against rain and waves are made by a 50 × 3 mm stainless steel flat profile: the subgrade is smoothed with mortar/shotcrete, a rubber gasket is placed on the even surface, the geocomposite is placed on the gasket and anchored with mechanical anchors.

In masonry dams, the roughness of the upstream face requires some modifications to the system used in concrete dams, such as placing under the waterproofing geo-composite an anti-puncture layer (Fig. 7 at left), at Kadamparai a 2000 g/m^2 geotextile, and levelling the masonry surface under the watertight peripheral seals with shotcrete and mortar. In case of extremely rough masonry, which was not the case at Kadam-parai, a levelling layer may be needed also under the tensioning profiles.

Fig. 7. Kadamparai: At left, the white anti-puncture geotextile placed on the masonry and the grey waterproofing geocomposite deployed over it. At right, the dam impounding after waterproofing works have been completed

The drainage system is divided into two top compartments, one at each side of the scour vent tower, collecting the water from the upstream face, and in two bottom compartments, one at each side of the scour vent tower, collecting the water between the primary and secondary seals. The bottom collector consists of a longitudinal band of highly transmissive geonet that conveys water to the discharge pipes to the gallery, where the drained water is measured. Additional monitoring systems consist of two piezometers, one for each of the upper compartment, to ascertain if there is water standing behind the geocomposite, and of an Optical Fibre Cable system, allowing locating the area of a possible leak. The system was developed by the Technical University of Munich in Germany and is based on the difference in temperature that is caused by leaking water and is signalled by the optical fibre cable. The optical fibre cable is placed and fastened on the masonry, in specifically designed loops terminating at a control box placed at crest.

The waterproofing works, for a total of 17,300 m^2 of surface lined, were completed in 10 weeks, 6 weeks ahead of schedule, which allowed earlier power generation. The owner (Sadagopan and Kolappan 2005) reports "The installation of the exposed PVC geocomposite mechanically anchored and drained has more than confirmed the expectations… Seepage that was in the order of 3×10^4 l/min has been reduced to around 1×10^2 l/min. The overall costs have been kept within the quoted price, and installation of the system has been completed within a very short time, ahead of schedule, allowing for the generation of power much earlier than expected. A geomembrane system, adequately designed and installed, is an advantageous alternative to traditional repair systems in terms of technical and financial effectiveness".

After 11 years of operation of the system, the seepage through the inspection gallery recorded on 26.11.2016, when the water level was 1148.30 m (F.R.L.: 1149.00 m), was 29 l/min for the geomembrane lined area (Fig. 8 at right). The advantages of a geomembrane system for rehabilitation of Kadamparai dam are the dramatic seepage reduction, the costs savings for maintenance of the dam (no more repeated local repairs), and the benefits of higher energy production. The same system is in 2018 under installation at Servalar 60 m high masonry dam, also in Tamil Nadu, which is the first dam to be rehabilitated under the DRIP - Dam Rehabilitation and

Improvement Project, launched by the Central Water Commission of India with assistance from the World Bank.

Fig. 8. At left, the inspection gallery of Kadamparai dam before waterproofing works. At right, the inspection gallery after waterproofing works: from 38,000 to 29 l/min

3.3   Asphalt Concrete Facing Fill Dams: Morávka, Czech Republic

In dams with asphalt concrete facing the tensioning system is generally used for face anchorage. Depending on the pull-out strength of the facing, the anchor rods + chemical phials used in concrete and masonry dams, as well as in CFRDs, are replaced by anchor rods + resins or by deep grouted anchors. Regarding perimeter seals, if they are made on a concrete plinth they are of the tie-down type already described, otherwise the insert-type seal is used, as done at several dams of this type and as discussed below.

Morávka dam was built in the 1960ies as a multiple purpose reservoir with the main scope of providing drinking water for the population in the Northern Moravian region, and the additional scopes of flood mitigation and power production. The dam, built as an earthfill dam with a multi-layer asphalt concrete facing with a total area more than 25,000 m^2 connected to the grouting gallery, is 39 m high and 396 m long at crest. The dam body is made of unsorted gravel-sand. The upstream slope has a 1:1.75 inclination.

The lack of experience and the shortage of proper material for sealing led to insufficient watertightness of the waterproofing facing. Morávka experienced leakage since the first uncontrolled filling in 1965; the reservoir was drawn-down, mastics cover and two new bituminous concrete layers were added to the original ones. At first seepage was reduced, but over time it increased again and despite several repairs defects and seepage continued until in 1997, when the biggest flood in the century caused the water level in the Morávka reservoir to rise to the maximum level. Evaluation of the data collected in that occasion showed that the main cause of the failure of the asphalt sealing had been the long-term erosion of the subgrade at the left bank of the valley, caused by the subsurface water percolating from the left bank and resulting in progressive enlargement of caverns near the surface asphalt sealing.

Povodí Odry, owner of the dam, decided to perform a total rehabilitation of the dam including also re-facing. The technical requirements for the new upstream sealing were proven long-term resistance to changeable weather conditions, frequent freeze/thaw cycles, ice formation, high temperature excursions, and the possibility of monitoring the performance on a continuous basis. The new sealing would have to deal with deformations and depressions in the subgrade. The length of realization of the new sealing, and the safety of the dam body during possible flood in the time of construction, were rated very high in the analysis. Povodí Odry prepared a cost-benefit analysis where the most important subjects and risks were considered. Adopting a concrete re-facing, and local repair of the existing surface only, were discarded since the beginning. The only solutions taken into consideration were a new asphalt concrete facing, and a geomembrane (exposed or covered) facing. The technology of removing the old and placing a new asphalt concrete facing was time consuming, riskier from the point of view of protection of the dam body during floods, and more expensive. After considering all possible risks and the time required for implementing it, an exposed geomembrane sealing system was selected. The time established for the installation in summer 1999 was 90 calendar days, which included 15 days as estimated time for stoppage due to floods and bad weather. An exposed geomembrane system was the only waterproof facing that could meet the technical requirements and the time constraints.

The waterproofing liner is the same of Kadamparai. Face anchorage is provided by the tensioning system already described; the asphalt concrete layer was tested and found to have sufficient pull-out strength to allow adopting anchor rods embedded in resin. The drainage system consists of the gap created by the anchorage system between the existing asphalt concrete facing and the new geocomposite facing. Drained water travels by gravity in this gap to reach a drainage collector placed along the bottom periphery and along the compluvium between the left bank and the straight part of the dam. The collector is a band of geonet with high in-plane transmissivity, positioned under the geocomposite.

At the bottom periphery there is a double watertight seal, the secondary one placed on the concrete plinth and of the tie-down type already discussed, and the primary one placed at the bottom of the asphalt concrete facing. The primary seal is of the insert type: a slot is made in the asphalt concrete, a geocomposite strip is inserted in the slot that is then filled with watertight resin (Fig. 9 at left), and the geocomposite waterproofing the upstream face is watertight seamed to the inserted strip. This type of seal has given excellent performance in dams with high water heads.

Fig. 9. Morávka: At left, preparation of the insert-type seal; at middle, Morávka dam in spring 2007; at right, during the visit of a delegation in 2008. Maximum total seepage at the dam is 1.0 l/s

The drainage system is divided into 11 compartments, each in turn further divided into a primary compartment draining water from the upstream face, and a secondary compartment draining water coming from foundation. Each compartment has an individual drainage discharge pipe, placed in a hole drilled to reach the inspection gallery, and equipped with a monitoring system.

When the 10 years guarantee was going to the end, the owner of the dam decided to make a final test of the sealing system. The reservoir was filled up to the emergency spillway during snow melting in spring 2009. The owner (Kratochvil et al. 2011) reports "After the stabilization of the measured values it was proved that the function of the upstream geomembrane sealing system is excellent. The value of seepage was under guarantee limits for water level on the spillway and attained a maximum value of 0.15 l/s from one section and 1.0 l/s in total, well below the minimum acceptable contract leakage of 2.0 l/s for the entire face".

3.4   Fill Dams: Vaité, Tahiti, French Polynesia

If the characteristics of the subgrade (granular materials, hardfill, asphalt facings with low pull-out strength) do not allow using anchor rods + chemical phials or resins, deep anchors can be used. Deep anchors can be of the "duckbill" type, as adopted at Vaité dam, or of the grouted type, as adopted in hardfill dams and at several hydropower canals.

Vaité is a 23 m high earthfill dam used for hydropower. In 1987, the upstream face of the dam was lined with a thin (1 mm) geomembrane placed on an anti-puncture geotextile, and anchored at top to the parapet wall, and at the abutments in a trench. The geomembrane was extended on the bottom of the reservoir for about 50 m, and on the abutments, so to provide some waterproofing of the foundations. In the years 2010–2011 this geomembrane was no longer functional and the owner of the dam Manama Nui-Electricité de Tahiti decided to substitute it. A requirement for the new geomembrane system was to resist 204 km/h winds (hurricane conditions) in the top 1/5 of the dam, and 100 km/h winds in the remaining 4/5. The waterproofing geocomposite was fastened to the dam face and at the abutments with duckbill anchors placed at depth, pattern and spacing designed to resist the design winds with the required safety factors. The system (patent pending) exploits the technology of deep anchors designed for anchorage in gravel soil, and consists of a stainless-steel tendon and of a rotating duckbill, covered by a suitable capping that guarantees that no water infiltration occurs where the anchor crosses the geocomposite, and that the forces at each anchor are adequately distributed so as to not overstress the geocomposite. The tendon and duckbill are driven into the ground and after the duckbill is in place, an upward pull on the tendon rotates it into a perpendicular "anchor lock" position in the soil that mobilises the resistance of the soil itself. See Fig. 10. In the six years that have elapsed since installation, the system has been performing as expected.

Fig. 10. Vaité: At left, the deep anchors can be seen protruding from the upstream face, before permanent fastening and waterproofing with a geomembrane element heat-seamed to the waterproofing liner. At right, the dam back in service

4 Waterproofing of Crucial Areas

The systems described for full-face rehabilitation can be used for repair of most leaking areas, as well as for staged rehabilitation to meet the financial needs of the owners.

An example of waterproofing of crucial areas is Grindstone Canyon RCC dam in USA. The dam, completed in 1986 and 42.3 m high, is used for water supply and owned by a Municipality. Seepage had been occurring at the dam since first filling. To mitigate seepage, the designers selected, as preferred alternative for a cost effective and viable waterproofing solution, an exposed geomembrane system installed over approximately the upper half of the dam. The system is like the one adopted at Kadamparai: the geocomposite is placed on an anti-puncture geotextile and anchored by vertical tensioning profiles. The challenges were installation, to be made with reservoir's level just below the area to be waterproofed (Fig. 11), and the porous dam body and fractured upstream face that could have significant interconnected seepage paths across the bottom seal. A urethane grout curtain was installed to cut off potential seepage paths around the bottom seal; cracks at the base of vertical construction joints, and the crossings joints/bottom seal, were grouted to cut off potential seepage paths.

Fig. 11. Grindstone Canyon: At left, the geocomposite under installation on the anti-puncture geotextile. At right, waterproofing works almost completed

Grindstone Canyon is one of the RCC dams that have used a geomembrane system as repair measure. The first use of geomembranes in RCC dams dates back to 1990, when they were used for new construction. Since year 2000 these dams, whose advent dates back to the beginning of the 1980ies, started needing repairs, which have been performed with geomembrane systems in the dry, and underwater.

5 Local Repair

In case repair is needed for cracks/failing joints, or for small local defects, Carpi patented external waterstop is placed over the failing area. The waterstop is a multi-layered system exposed to the water of the reservoir, and sealed along the perimeter by a watertight mechanical seal of the type already described. Depending on the size of the defect, on the expected opening of the joint, on the dimensions of crack/hole, on the water head, and on the conditions of the subgrade, one or more layers may be needed under the waterproofing liner. Such layers solve different functions:

(a) Anti-puncture/sacrifice, to protect against aggressive subgrade - generally, a non-woven geotextile or a geocomposite
(b) Support, to impede that the waterproofing liner intrudes in the active joint/large crack and, in case of very large openings and of high water head, that it bursts - generally, this function is solved by one or more layers of flexible material, with lower or higher modulus depending on the needs, seldom by a rigid support
(c) Anti-friction, to avoid the waterproofing liner being affected by the movements of the underlayers - generally, a flexible layer
(d) Waterproofing: the geocomposite.

The waterproofing geocomposite, typically 40 to 70 cm wide, is anchored so that it can elongate over its entire width. The geotextile having an elongation at break around 60%, and the geomembrane elongation at break exceeding 250%, openings with order of magnitude of tens of centimetres/exceeding 1 m can be accommodated before respectively the geotextile and then the geomembrane break.

Concrete dams, CFRDs, RCC dams, intake structures of asphalt concrete facing dams and reservoirs, have been rehabilitated with this external waterstop. Figure 12 shows a typical layering and an example of dry installation at Usina da Pedra buttress dam in Brazil, Fig. 13 an example of underwater installation at the failing joints of Olai concrete gravity dam in Italy. The same system has been adopted for new construction, on contraction joints of RCC dams and on peripheral + vertical joints of CFRDs, where it was designed for openings up to 30 cm.

Fig. 12. Carpi external waterstop: a typical layering (contraction joints at Platanovryssi 95 m high RCC dam), and ongoing dry installation at Usina da Pedra buttress dam

Fig. 13. Placing underwater the external waterstop on three failing joints at Olai reduced leakage from 65 l/s to 0.33 l/s

6 Underwater Installation

Possibility of underwater installation is undoubtedly a very attractive feature of geomembrane systems, because it requires minimum or no disruption in operation of the dam during the rehabilitation works. In its second phase, the above mentioned two-years research by USACE developed a geomembrane system that could be installed underwater, and was tested in full-scale on a reinforced concrete struc-ture 4.8 m × 3 × 1.5 m replicating different features that can be encountered in underwater installations, such as subgrades of different roughness, joints, holes, complicated geometries. The structured was lowered in a tank where the geomembrane system was installed at about 6 m depth. A suction corresponding to 8.4 m of hydrostatic head was applied to the back of the geocomposite liner for 2 weeks. The geocomposite conformed to the subgrade, and no leaks were detected, demonstrating that an effective seal had been achieved (Christensen et al. 1996). The research demonstrated the feasibility of underwater installation, and allowed to refine design of the various components of the system, and of underwater installation procedures. After successful full-scale testing of the system conceived in the second phase of the research, the system was validated for underwater installation.

The first underwater installation was a full-face repair made in 1997 in USA, at Lost Creek arch dam. The first example of underwater installation in staged campaigns following the funding available to the owner was Turimiquire (Las Canalitas) 113 m high CFRD in Venezuela, owned by the Ministerio del Poder Popular para el Ambiente and used for potable water supply. Already one year after impoundment, in 1988, the dam showed 300 l/s leakage, which, despite regular repair works carried out over time, eventually increased up to 9,800 l/s. Underwater investigation ascertained that the main causes of leakage were cracks and two craters in the concrete slabs. Rehabilitation works were carried out in 2010/2011 in the most critical areas, at water depth exceeding 50 m. The waterproofing liner is a 3 mm thick geomembrane heat-bonded to a 700 g/m^2 geotextile. The fastening system for underwater installation is the tensioning system already described, slightly modified to adapt it to the underwater working environment. A support geogrid was placed on the concrete face at largest discontinuities, and a 2000 g/m^2 anti-puncture geotextile was applied on the entire area, under the waterproofing geocomposite. The first-stage rehabilitation works covered about 1/5 of the dam face, and allowed decreasing leakage to 2,400 l/s (Fig. 14 at right), well below the target leakage of 3,000 l/s. This helped to maintain the dam in service and, in the meantime, to gather new funds for continuing the rehabilitation works. In 2016/2017, the second stage was performed in the area of the second crater, covering an additional 1/20 of the upstream face.

Fig. 14. Turimiquire CFRD with (in white) the geomembrane system during first-stage underwater waterproofing works. At right, the reduction in water leakage after waterproofing about 1/5 of the dam upstream face

In 2017/2018, a full-face underwater rehabilitation is ongoing at Studena 55 m high buttress dam in Bulgaria, which is used for potable water supply and where an improved version of the tensioning system is being adopted (Fig. 15). Studena also is a World Bank project.

Fig. 15. Underwater waterproofing works ongoing at Studena buttress dam

7   Conclusions

Lining systems with the geocomposites presented in this paper provide durable continuous watertightness and can re-face traditional rigid liners, providing technical and financial benefits. They can bridge joints and large fissures, resist opening of new fissures in case of settlements in the natural slopes, and achieve seepage rates typically extremely low. The flexibility and robustness of the geocomposite allows it adapting to complicated geometries and rough subgrades. Installation times and costs are lower than those of other rehabilitation systems, practically no maintenance is required, and outstanding durability has been proven.

References

Christensen, J.C., et al.: A conceptual design for underwater installation of geomembrane systems on concrete hydraulic structures. Technical Report REMR-CS-50 (1995)

Christensen, J.C., et al.: A constructibility demonstration of geomembrane systems installed underwater on concrete hydraulic structures. Technical Report REMR-CS-51 (1996)

Giroud, J.P.: Analysis of stresses and elongations in geomembranes. In: International conference on geomembranes, vol. 2. IFAI Publisher (1984)

ICOLD, The International Commission on Large Dams: Bulletin 117: The gravity dam – a dam for the future – review and recommendations (2000)

ICOLD, The International Commission on Large Dams: Bulletin 135: geomembrane sealing systems for dams – design principles and review of experience (2011)

Kratochvil, D., et al.: Exposed geomembrane system at Morávka dam: 10 years performance. In: Hydro 2011: Practical Solutions for a Sustainable Future (2011)

Peggs, I., et al.: Assessment of maximum allowable strains in polyethylene and polypropylene geomembranes, geo-frontiers congress (2005)

Sadagopan, A.A., Kolappan, V.P.: Rehabilitation of Kadamparai dam, India. Int. J. Hydropower Dams **12**(4), 79–81 (2005)

Scuero, A., Vaschetti, G.: Geomembrane sealing systems for dams—ICOLD Bulletin 135. In: GeoMEast 2017 International Congress (2017)

Long Term Behavior of EPS Geofoam for Road Embankments

Sherif S. AbdelSalam[1]([✉]), Mona B. Anwar[2], and Sylvia S. Eskander[2]

[1] Civil and Infrastructure Engineering and Management,
Faculty of Engineering and Applied Science, Nile University, Giza 12588, Egypt
sabdelsalam@nu.edu.eg
[2] Civil Engineering, Faculty of Engineering and Materials Science,
German University in Cairo, Cairo, Egypt
{mona.anwar, sylvia.emil}@guc.edu.eg

Abstract. In recent years expanded polystyrene (EPS) geofoam has success-fully been used to reduce the acting vertical and horizontal stresses in several geotechnical applications due to its light weight, compressibility, and durability. In this study the efficiency of utilizing EPS blocks as a replacement for typical soil embankments under roadways was investigated. Accordingly, a detailed laboratory program was completed to measure the short- and long-term behaviors of EPS, which included unconfined compression (UC) and creep strain (CS) tests based on the time–temperature–stress superposition (TTSS) technique. Loading type applied during testing was cyclic loading to mimic the actual conditions under roadways, while the EPS density used was 35 kg/m^3 to minimize deformations. The main outcome of this study was providing mea-sured properties for local EPS considering creep stain after 100 years of cyclic loading, and these properties are ought to provide reliable design for EPS embankments under roadways.

1 EPS for Road Embankments

Expanded polystyrene (EPS), or geofoam has been used in geotechnical applications since the 1960's. Reference to EPS Industry Alliance (2012), EPS is approximately 1% of the soil weight, and is less than 10% the weight of other lightweight fill alternatives. This makes EPS embankments lighter compared with compacted soil embankments, and accordingly reduces the dead loads imposed on underlying soil. This also translates into benefits to construction cost and time, and simplicity of construction as EPS blocks are easy to handle without the need for special equipment, and can be easily cut and shaped on site. EPS geofoam is available in various densities that can be selected by the designer depending on a specific geotechnical application. Its durability and service life are comparable to other construction materials as it retains its physical properties under various conditions (AbdelSalam and Azzam 2016).

In 2012, Bartlett et al. (2012) studied the used EPS as an embankment for the interstate roadway I-15 in Salt Lake City, Utah, USA. The idea of using EPS blocks was to mitigate settlement and expedite construction on soft clay, while the importance of using EPS was to improve the stability of the embankment. At some bridge

© Springer Nature Switzerland AG 2019
M. Meguid et al. (Eds.): GeoMEast 2018, SUCI, pp. 97–107, 2019.
https://doi.org/10.1007/978-3-030-01944-0_8

locations, high embankments were required and the calculated safety factors against base failure were low. Those embankments are usually constructed with geotextile reinforcement and staged embankment construction that require several months to allow excess pore pressure dissipation and subsequent shear strength gain. Construction of embankments with EPS blocks provided higher safety factors against geotechnical instability and reduced the construction time by about 75%.

In order to use EPS in geotechnical applications, especially for road embankments, there are several parameters that should be determined such as the EPS density, axial compressive strength, effect of creep on the long-term, and also the effect of cyclic loading. The density of EPS is controlled through regulations of the manufacturer, and it has been found that even with a well-controlled manufacturing process there will still be variability in density between blocks from the same production run, as well as a density gradient within each block (Lutenegger and Ciufetti 2009). According to Horvath (2004), this variation may affect the geotechnical design properties of the material, as density is a controlling factor. Internationally, the actual range of densities that EPS blocks can be manufactured is between approximately 20 kg/m^3 and 40 kg/m^3. Locally, the average density of the most commonly produced EPS blocks is around 30 kg/m^3.

Another factor that controls the EPS behavior is the axial compressive strength. The EPS compressive behavior can be determined by testing cubes with dimensions 50 mm × 50 mm × 50 mm under strain-controlled unconfined axial loading at rates of up to 20% strain per minute, with 10% being the most common (Lutenegger and Ciufetti 2009). The stress–strain response of EPS can be divided into zones consisting of initial linear response, yielding, and post-yield nonlinear hardening. Several studies such as Meguid and Hussein (2017) have found that the initial linear-elastic portion of the curve extended up to 2% strain; however some researchers have reported slight variations in this range to reach up to 5%. It has now been commonly accepted that the elastic limit stress corresponds to a compressive strain between 2 to 5%, where slope of the linear-elastic portion of the curve is the initial modulus of elasticity (AbdelSalam et al. 2015).

As known that EPS is a polymer so it is considered a creep-sensitive material, which means that any sustained compressive load applied would result in compression creep (Koerner 2012). According to Yeo (2007), creep can be defined as time-dependent strain that could induce deformations in the structural system. Koerner (2012) mentioned that in designing with geosynthetics, the laboratory results must be reduced using reduction factors that represent site-specific conditions or by using a conservative factor of safety. To evaluate the creep strain behavior of EPS, a creep master curve should be derived utilizing either the stepped isothermal method (SIM), the time–temperature superposition (TTS), or the time–temperature–stress superposition (TTSS). Based on recommendations by Yeo (2007), the TTSS is an efficient method to determine the creep strain of EPS. TTSS is an accelerated creep test based on ASTM-D-2990 (ASTM 2005) that utilizes temperature and stress to accelerate creep strains. This method depends on determining creep properties at a reference temperature and stress corresponding to working conditions (Yeo 2007). The principle allows for equivalence of time–temperature and time–stress. Therefore, creep curves obtained under conditions different to the reference are superposed by horizontal shifting along a logarithmic scale to develop a master creep curve. Then, appropriate total creep strain can be selected from the master creep curve utilizing the design life cycle of the roadway. Yeo (2007) also indicated that a reduction factor must be applied to the ideal compressive strength to account for the creep deformation and incorporate that in the design.

According to Shi et al. (2016), EPS is considered as an energy-absorbing material for some specific constructions that undergo long-term cyclic loading such as road-ways. In the application of using EPS as road embankments, this requires conducting cyclic uniaxial compression (CUC) on EPS specimens and comparing the results with the static uniaxial compression to determine the number of cycles where the EPS strength starts to decline. The EPS behavior under cyclic loading, or the CUC can be determined by testing cubes with dimensions 50 mm × 50 mm × 50 mm and after applying the appropriate number of load cycles until no significant variation in the material properties is evident. The reduction that occurs in the EPS modulus of elasticity due to cyclic loading shall be considered in the design.

2   Laboratory Testing

As there is lack in the long-term properties of local EPS, a comprehensive laboratory testing program was planned and carried out in this study to characterize the main design properties of EPS with density (ρ) equals to 35 kg/m^3. This was achieved through implementing a series of tests such as the unconfined compression (UC) to determine the initial modulus of elasticity (E) and the compressive strength (σ_c), the modified direct shear test (DST) as per recommendations by AbdelSalam et al. (2017) to determine the shear strength parameters (cohesion, c, and angle of internal friction, ϕ), the creep strain test (CS) using the TTSS technique to assess creep strain after 100 years of load application, and finally the cyclic uniaxial compression test (CUC) to determine the reduction in E and σ_c with respect to number of load cycles.

2.1   EPS Short-Term Modulus of Elasticity

A series of static UC tests was applied on 0.35 kN/m^3 EPS cubical shaped blocks 50 × 50 × 50 mm with loading rate 5 mm/min corresponding to around 10% strain rate as per recommendations by AbdelSalam et al. (2015). Results of the UC test are shown in Fig. 1, where the measured stress–strain curves for four different EPS samples are represented including the average stress–strain curve. The equation of this average stress–strain curve is provided on the figure, which also shows that that the initial modulus E was equal to around 8000 kPa at strain 1%, while the compressive strength was around 215 kPa.

From Fig. 1 it can also be noticed the occurrence of failure zones as specified in the literature; firstly, the initial linear response (zone 1) which reached about 2.5% strain. Secondly, the yielding zone (zone 2) up to around 20% strain and began with a definite yield point to signify plastic deformation. Thirdly, the linear hardening zone (zone 3) where EPS specimens became increasingly stiffer. It is noteworthy to point out that

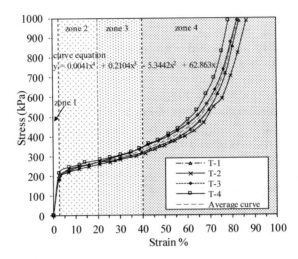

Fig. 1. Unconfined compression on four EPS samples (short-term)

strain past 40% began to create a positive slope on the curve, this is due hardening and would continue to increase into zone 4 (non-linear hardening zone).

2.2 Shear Strength Properties

The material used to produce EPS is originally composed of small beads with a diameter close to medium coarse sand, whereas these beads have internal shear strength (i.e., cohesion, c, and internal angle of friction, ϕ). Recently, Khan and Meguid (2018) determined the shear strength parameters of EPS using three different densities 15, 22 and 35 kg/m^3, and it was specified that there is a direct correlation between the EPS density and its cohesion, and an indirect correlation with its internal angle of friction. Khan and Meguid (2018) found that the EPS unit block experienced shear deformation with no clear failure surface, whereas the c = 54 kPa and ϕ = 9°.

Accordingly, a DST was conducted herein following similar procedures as stated by Xenaki and Athanasopoulos (2001) to determine the maximum shear strength of a unit block of local EPS. This test was conducted for a block occupying a squared DST with box dimensions 60 × 60 × 30 mm. The normal stresses applied during testing were 10, 30 and 40 kPa, and horizontal displacement up to 9 mm was enforced to cover potential phases of shear failure within each loading stage. The test was conducted in dry conditions following the ASTM standards (ASTM D 3080). Figure 2a represents shear stress and horizontal displacement for unit EPS block under different normal stresses in dry conditions. As shown in Fig. 2a, the shear stress increased when increasing the normal stress, which follows the typical behavior of medium dense soil. Figure 2b represents the shear failure envelop, where the cohesion of the EPS beads was about 55 kPa and the ϕ was about 15.2°. These measured results are quite close to what was indicated by Khan and Meguid (2018).

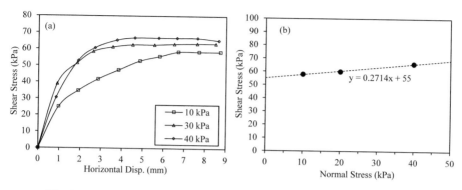

Fig. 2. Direct shear results (a) shear stress–displacement; (b) failure envelop

2.3   EPS Long-Term Creep Strain

Creep is defined as time-dependent strain that induce secondary or tertiary deformations in a system, which represent the difference between short- and long-term
behaviors of EPS. Following normal procedures for the TTSS test, the specimens
should be brought to equilibrium prior to testing by placing them in a temperature
regulated water chamber (TRWC); however, that TRWC was not available and
therefore the Oedometer cell was slightly modified and utilized as a simple alternative
for TRWC. In the Oedometer cell, the vertical stresses can be applied on the EPS
sample and maintained whilst having a chamber that can be filled with hot water. The
Oedometer cell was covered from top to preserve the water temperature inside the
chamber, meanwhile a digital thermometer was used during the entire testing period.
The water inside the Oedometer cell was maintained at ±2 °C of testing temperature as
per the ASTM-D-2990 standards (ASTM 2005).

In the creep strain (CS) test, an EPS cylindrical specimens with diameter 50 mm
and height 20 mm was used to fit the loading ring of the Odometer. Before inserting the
specimens in the odometer, all EPS samples were soaked in water for 12 h at room
temperatures. Then samples were placed in the Odometer for 4 h under normal stresses
of 15%σ_c, 25%σ_c, and 40%σ_c. Each stress range was tested in temperatures equal to
23 °C, 33 °C, and 43 °C, whereas these temperatures were chosen according to Yeo
(2007) knowing that the temperature did not exceed 44 °C to prevent a premature
secondary creep stage. Hence, the total number of CS tests was nine tests on EPS with
density 35 kg/cm³. Figure 3a–c represent CS results at various temperatures obtained
for samples subjected to normal stresses of 15%σ_c, 25%σ_c, and 40%σ_c, respectively.
As expected, the results showed that the creep strain increase by increasing the temperature and acting stress. For instance, the maximum creep strain at a normal stress of
15%σ_c was only about 0.7% at 23 °C, while creep stain considerably increased to
exceed 4.0% at a normal stress of 40%σ_c and 43 °C.

To create the master creep curve, the first step was to plot the CS results using log-
time scale as presented in Fig. 4. The resulting creep curves illustrate a creep strain
increase with respect to time, applied stress, and temperature, whereas for the nine tests

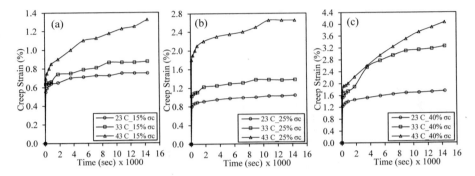

Fig. 3. Creep strain test results at stresses (a) 15σc; (b) 25σc; and (c) 40σc

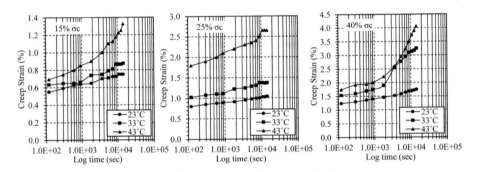

Fig. 4. Creep strain vs. log time for EPS 35 at various stress and temp. ranges

the creep strain was retained within the primary creep zone as expected. The creep curves were then superimposed by horizontal shifts following the TTSS method to generate the master creep curve as shown in Fig. 5. The developed master creep curves presented in Fig. 5 cover a period of time that exceeded 100 years, which satisfy the maximum road design life cycle. The obtained creep strains of EPS with density 35 kg/m³ at 100 years was around 2.0%. This means that the original stress–strain curve of EPS 35 (i.e., EPS with density = 35 kg/m³) that was previously shown in Fig. 1 shall be multiplied by a strain reduction factor of 2% to account for creep strain after 100 years of service.

2.4   Cyclic Loading Effect on EPS Behavior

The cyclic uniaxial compression (CUC) test were carried out on $50 \times 50 \times 50$ mm cubes to mimic the actual conditions under roadways at frequency equals to 1% strain as the design of roadways has to be in the linear elastic zone with stress range from 0 to 100.44 kN/m². The number of manual load cycles used in this test was 100 in order to monitor the reduction in the material properties. As can be seen in Fig. 6a, the CUC

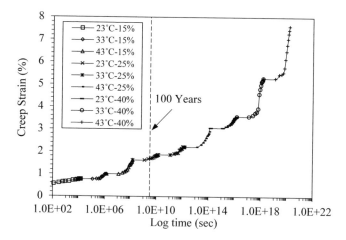

Fig. 5. Master creep curve using TTSS superposition for EPS 35

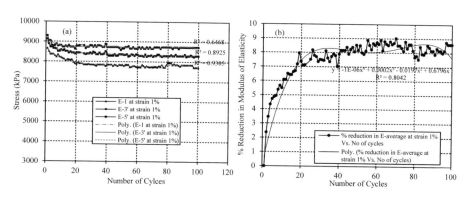

Fig. 6. Effect of CUC on EPS 35 (a) stress vs. cycles; and (b) reduction in E

was conducted on three different EPS samples with same density of 35 kg/m³. The three samples were used in order to make sure that the test is replicable with acceptable accuracy. As shown from the figure, results were subject to some differences that did not exceed 5 to 7%, also it is obvious that effect of load cycles on the material behavior was high within the first 20 cycles, then this effect started to be negligible after 50 cycles. Figure 6b represents the percentage of reduction in the modulus of elasticity, E, versus the number of load cycles, which also includes the reduction curve equation. From this figure, it can be noticed that the maximum reduction in E was around 8%, which shall also be considered in the design.

3   Numerical Modeling

The main aim of the numerical model is to confirm that the measured behavior of EPS under axial compression can be simulated using the hardening soil (HS) constitutive model of Plaxis 2D v.17 (Brinkgreve et al. 2015). The modulus of elasticity used in the model was based on laboratory measurements and after reduction to account for creep and cyclic loading. To confirm that the HS can capture the accurate behavior, a calibration factor was introduced for the modulus of elasticity to be used in Plaxis to adjust the calculated with the measured responses.

3.1   Stiffness and Material Modulus

Finite element analysis of EPS under unconfined compression test was carried out using an axisymmetric model. The initial model is shown in Fig. 7a, where model EPS diameter was 25 mm and height 50 mm. The vertical boundaries of the model were released in the vertical direction, and the bottom boundary was totally fixed. A 15-nodded unstructured mesh was used as shown in Fig. 7a, and the loads were assigned using five loading stages starting from 215 kPa up to 1075 kPa. A steel plate was used on top of the EPS sample to evenly distribute the stress on the sample. As previously mentioned, the constitutive model used for the EPS was the HS model as per recommendations by Horvath (1994), Awol (2012), and AbdelSalam and Azzam (2016), with the parameters provided in Table 1. The only derived parameter was the Poisson's ratio (υ), which was based on the equation provided by Edo (1992) as follows: $\nu = 0.0056\,\rho + 0.0024$. Figure 7b provides the mean shading for vertical displacement in the y-direction, while Fig. 7c shows the effective normal stresses induced in the sample during the last phase of loading. It is important to highlight that all the EPS parameters used in the model were based on the primary lab results including the creep and cyclic loading reduction effects, appearing in the modulus of elasticity (i.e., before calibration).

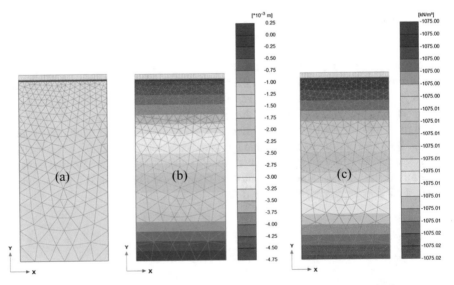

Fig. 7. Axisymmetric model for EPS (a) mesh: (b) disp in y-dir; (c) vertical stress

Table 1. Material parameters used in the numerical model

Material	EPS Geofoam	Steel Plate
Model	Hardening Soil (HS)	Elastic
γ (kN/m$^{3)}$	0.35	N/A
Cohesion, C (KPa)	55	N/A
Friction angle, ϕ (°)	15	N/A
Dilatancy angle, ψ (°)	0	N/A
Initial stiffness, E_{ref} (KPa)	8000	196.3E3[a]
Secant stiffness, E_{50} (KPa)	8000	0.01023[b]
Poisson's ratio, ν	0.2	0.25
Calibration factor	1.3125	N/A

[a]EA for steel plate; where E = young's modulus; and A = area of plate
[b]EI for steel plate; where I = inertia of plate.

For a point chosen at the middle on the centerline of the model, the relation between the stress and strain was calculated and it was found that there is a difference of about 30% between the slope of the calculated and measured curves. This could be due differences between the actual EPS sample and model, also the program did not capture the effect of creep and cyclic loading although embedding them in the modulus of elasticity. Hence, the material parameters, mainly the modulus of elasticity, was multiplied by a correction factor in an attempt to match the actual measured behavior of the material. A factor of 1.3125 was acquired after several iterations as indicated in Table 1. Figure 8 represents the measured stress–strain curves for EPS before considering creep and cyclic effects, and after considering these effects. Modulus of elasticity used in the HS constitutive model of Plaxis 2D was calibrated based on results shown in Fig. 8.

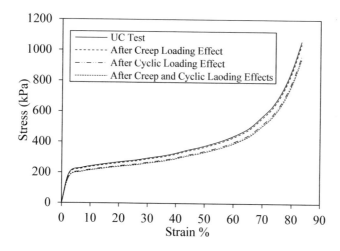

Fig. 8. EPS stress–strain curves before and after creep and cyclic effects

4 Conclusions

Expanded polystyrene (EPS), or geofoam is used now in many geotechnical applications as its weight is approximately 1% of the soil weight and is less than 10% the weight of other lightweight fill materials. This makes EPS embankments relatively light compared with compacted soil embankments, and accordingly reduces the own weight imposed on underlying soils. In roadways applications, the long-term properties of EPS are essential such as creep and cyclic loading, and there is no enough information about these parameters in the existing literature. In an attempt to determine the long-term properties of EPS, a comprehensive laboratory testing program was carried out to characterize the main design properties of EPS with density, $\rho = 35$ kg/m^3. A series of tests such as the unconfined compression (UC), direct shear test (DST), creep strain test (CS), and cyclic uniaxial compression (CUC) were carried out. The major outcomes of this experimental program are summarized as follows:

1. Short-term stress–strain of the local EPS 35 was linear elastic until a compressive strain of about 2.5%.
2. Shear strength parameters for a unit block of EPS 35 were: cohesion, $c = 55$ kPa; and internal friction angle, $\phi = 15.2°$.
3. Creep strain increased by the factors of time, applied stresses and temperature.
4. The creep strain of local EPS 35 at 100 years was found to be around 2%.
5. EPS modulus of elasticity was inversely proportional to number of load cycles.
6. Cyclic loading reduced the EPS modulus of elasticity by 8% after 100 years.
7. The hardening soil constitutive model provided a stress–strain curve for EPS 35 under axial loading with a difference of about 30% compared with the measured response in the long-term.
8. Calibration factor of 1.313 was multiplied by the modulus of elasticity used in the hardening soil constitutive model to adjust the model outcomes.

References

AbdelSalam, S.S., Azzam, S.A.: Reduction of lateral pressures on retaining walls using geofoam inclusion. ICE Geosynth. Int. **23**(6), 395–407 (2016)

AbdelSalam, S.S., Azzam, S.A., Abdel-Awad, S.A.: Laboratory characterization and numerical modeling of EPS geofoam. In: International Conference on Advances in Structural and Geotechnical Engineering, Hurghada, Egypt (2015)

AbdelSalam, S.S., Azzam, S.A., Fakhry, B.M.: Reliability and 3D modeling of flexible walls with EPS inclusion. ASCE Int. J. Geomech. **17**(7) (2017). https://doi.org/10.1061/(ASCE)GM.1943-5622.0000853

ASTM D 2990: Standard test methods for tensile, compressive, and flexural creep and creep rupture of plastics. ASTM International, West Conshohocken (2005)

ASTM D 3080: Standard Test Method for Direct Shear Test of Soils Under Consolidated Drained Conditions, Annual Book of ASTM Standards. ASTM International, West Conshohocken, PA

Awol, T.: A Parametric Study of Creep on EPS Geofoam Embankments. Department of Civil and Transport Engineering, Norwegian University of Science and Technology, Trondheim (2012)

Brinkgreve, R.B.J., Kumarswamy, S., Swolfs, W.M.: Plaxis 3D Reference Manual Anniversary Edition Version 1. Plaxis, Delft (2015). ISBN 978-90-76016-19-2

EDO: Expanded Polystyrene Construction Method. R. Tosho Pub., Tokyo (1992)

Horvath, J.S.: Expanded Polystyrene (EPS) Geofoam: An Introduction to Material Behavior. Elsevier, London (1994)

Horvath, J.S.: Lessons learned from failure: EPS geofoam. Geotechnical Fabrics Report, November, 2004, pp. 34–37 (2004)

Khan, M.I., Meguid, M.A.: Experimental investigation of the shear behavior of EPS geofoam. Int. J. Geosynth. Ground Eng. (2018). https://doi.org/10.1007/s40891-018-0129-7

Koerner, R.M.: Designing with Geosynthetics, vol. 2, 6th edn. Xlibris, Bloomington (2012)

Lutenegger, A.J., Ciufetti, M.: Full-scale pilot study to reduce lateral stresses in retaining structures using geofoam. Final Report, Project No. RSCH010-983 Vermont DOT, University of Massachusetts, Amherst, MA (2009)

Meguid, M.A., Hussein, M.: A numerical procedure for the assessment of contact pressures on buried structures overlain by EPS geofoam inclusion. Int. J. Geosynth. Ground Eng. 3(2), 1–14 (2017). https://doi.org/10.1007/s40891-016-0078-y

Bartlett, S.F., Negussey, D., Kimble, M.: Design and evaluation of expanded polysteren geofoam embankments for the I-15 reconstruction project. Utah Department of Transportation Research Division, Salt Lake City (2012)

The EPS Industry Alliance: Expanded polystyrene (EPS) geofoam applications and technical data (2012)

Shi, W., Miao, L., Luo, J., Wang, J., Chen, Y.: Durability of modified expanded polystyrene concrete after dynamic cyclic loading. Shock Vib (2016). https://doi.org/10.1155/2016/2391476

Xenaki, V.C., Athanasopoulos, G.A.: Experimental investigation of the interaction mechanism at the EPS geofoam-sand interface by direct shear testing. Geosynth. Int. 8(6), 471–499 (2001)

Yeo, S.S.: Evaluation of Creep Behavior of Geosynthetics Using Accelerated and Conventional Methods. Drexel University, Philadelphia (2007)

Flexural and Shear Characterization of Geosynthetic Reinforced Asphalt Overlays

V. Vinay Kumar[✉] and Sireesh Saride

Department of Civil Engineering, IIT Hyderabad, Kandi, Telangana, India
christite.vinay@gmail.com, sireesh@iith.ac.in

Abstract. In the current study, the flexural and shear characteristics of geosynthetic-reinforced asphalt overlays placed on a pre-cracked old pavement layer are evaluated in two different stages. The unreinforced and geosynthetic reinforced two-layered asphalt beam specimens with 40 mm deep notch (crack) in the bottom layer are tested under repeated four-point bending test and interface shear strength test equipments to understand the flexural and shear characteristics, respectively. The two-layered asphalt specimen consists of a 45 mm thick old pavement layer extruded from an existing highway as a bottom layer, a binder tack coat, an interlayer and finally a hot mix asphalt (HMA) overlay. Two different types of geosynthetic-interlayers, namely, biaxial polyester grid coated with polymer modified binder having a square aperture of 18 mm (PET) and a glass-grid composite (GGC) are used in the study.

The flexural fatigue and interface shear strength test results indicate that the reinforced specimens improved the fatigue life of overlays and among them, the performance of GGC specimens are superior. However, the inclusion of geosynthetic-interlayers at the interface of old and new layers reduce the interface shear strength, resulting in a possible delamination of the pavement layers. A reduction of 17% and 36% in interface shear strength was witnessed in PE, and GGC reinforced specimens, respectively. Overall, the geosynthetic-interlayers improved the performance life of asphalt overlays invariably, before failure.

Keywords: Geosynthetics · Asphalt overlay · Fatigue · Shear

1 Introduction and Background

Reflection cracking is termed as one of the major distress in the asphalt overlays across the globe by various pavement engineers and researchers. Reflection cracking can be defined as a phenomenon of propagation of cracks and discontinuities from the existing pavement layers into and through the hot mix asphalt (HMA) overlays (Cleveland et al. 2002; Saride and Kumar 2017). There exist innumerous factors that influence the rate of reflective cracking, and among them, the two driving factors are the repeated traffic wheel loads and the daily/seasonal temperature variations (Kim and Buttlar 2002; Khodaii et al. 2009; Kumar 2017). These cracks are observed to result in premature failures of the pavement system, as they allow the moisture ingression into the

© Springer Nature Switzerland AG 2019
M. Meguid et al. (Eds.): GeoMEast 2018, SUCI, pp. 108–123, 2019.
https://doi.org/10.1007/978-3-030-01944-0_9

pavement layers below through the cracks (Elseifi and Al-Qadi 2004; Kumar and Saride 2018a).

The reflection-cracking phenomenon is termed as a complex process, since innumerous factors influence the rate of reflective cracking, including the overlay thickness, traffic conditions, weather conditions, and the condition of existing old pavement layer. Hence, there exists no unique solution to arrest the reflection cracks completely. Besides, there exist few important solutions to retard the rate of reflective cracking and includes, providing a stress relief interlayer (Carey 1975), increasing the thickness of asphalt overlay (Sherman 1982), cut and seal technique (Al-Qadi et al. 2006), providing a chip-seal asphalt surface treatment (Gransberg 2006), and providing geosynthetic-interlayers (Kumar and Saride 2017; Saride and Kumar 2017). Researchers (Al-Qadi et al. 2006; Barraza et al. 2010; Kumar and Saride 2018b) worldwide have evaluated the effectiveness of various treatment techniques available to restrict the rate of reflective cracking. They have suggested that providing a geosynthetic-interlayer at the interface of old and new pavement layers is highly effective in resisting the crack propagation into the overlays. The inclusion of geosynthetics in the pavement systems have been increased drastically due to their ease in installation and their functions such as: providing a moisture barrier, reinforcement, separation, filtration and drainage (Khodaii et al. 2009).

Caltabiano (1990) performed a series of beam tests to study the performance of geogrids and fabric interlayers in restricting the crack growth into the overlays and found an improvement in the performance life of overlays reinforced with geosynthetic-interlayers. Similarly, Sanders et al. (1999) studied the crack retarding capacity of grids and textiles, and reported a two-fold improvement in the performance of grid-reinforced overlays, against cracking. Besides, Cleveland et al. (2002) studied the influence of geosynthetics on the performance of HMA overlays under Texas Transport Institute (TTI) Overlay tester and found that the grids were effective compared to fabrics and composites in resisting the crack growth. In addition, Walubita et al. (2015) attempted to study the cracking and fracture properties of the geosynthetic-reinforced asphalt layers using the Overlay tester and reported that the geosynthetic-interlayers have a potential to improve the fracture characteristics of asphalt layers.

Correia and Zornberg (2018) adopted the wheel-tracking test facility to study the strain distribution patterns in the geosynthetic-reinforced asphalt overlays. Based on the strain distribution patterns, they suggested that the geogrid-reinforcements provided in the asphalt overlays reduce the vertical and horizontal deformations in the asphalt layer. Virgili et al. (2009) and Ferrotti et al. (2012) adopted the flexural fatigue tests to study the influence of geosynthetics on the performance improvement of asphalt overlays. They found that the geosynthetics improved the performance by almost twice the control specimens. In addition, Kumar and Saride (2017) and Saride and Kumar (2017) performed beam fatigue tests with the aid of digital image correlation to understand the crack growth in the grid-reinforced overlay system. They reported that the presence of geosynthetic-interlayers at the interface of old and new pavement layers restricted the crack growth into the overlays and the cracks were redirected in the horizontal direction. Based on the observations made by Prieto et al. (2007), Saride and Kumar (2017), Wargo et al. (2017) and Kumar and Saride (2018a), the geosynthetic-interlayers not only improves the performance but may also reduce the bond strength between the pavement layers. Similarly, a reduction in shear resistance was witnessed with the inclusion of geosynthetics (Kumar et al. 2017; Saride and Kumar 2017).

Based on the recent-past research studies, it can be noted that the geosynthetics are mostly placed within the asphalt layers, instead of placing them at the interface of old and new pavement layers. In this regard, to address the actual behavior of geosynthetic-reinforced asphalt overlays, the current study considers a two-layered asphalt specimen with geosynthetic-interlayers placed at the interface of old and new pavement layers. The old pavement layer was extruded from an existing highway during a rehabilitation project and employed accordingly to replicate a real field scenario. The main objective of the study is to understand the flexural and shear characteristics of the geosynthetic-reinforced asphalt overlays using flexural fatigue and interface shear strength tests, respectively.

2 Materials and Methods

2.1 Geosynthetics

Two different types of geosynthetic-interlayers namely: bi-axial polyester grid coated with a polymer-modified binder (PET) and a glass-grid composite (GGC) were adopted based on their tensile strength characteristics, aperture size, bonding characteristics, and the material composition. Wide-width tensile strengths as per ASTM D4595 were performed along the machine direction (MD) and cross-machine direction (CMD) of the geosynthetic-interlayers to understand their working mechanical and tensile characteristics.

The bi-axial polyester grid is manufactured using a high molecular weight and high tenacity polyester yarns. The high tenacity polyester yarns are knitted to form a grid structure having a square aperture of 18 mm and a rib thickness of 3 mm and 4 mm along the MD and CMDs respectively, as shown in Fig. 1. The polyester grid is coated with a polymeric modified binder and has a thickness of about 2 mm. The grid has an ultimate tensile strength of 48 kN/m (MD) at a strain of 18%, and 52 kN/m (CMD) at a strain of 20%.

The composite interlayer comprises of a glass grid and continuous non-woven filaments mechanically bonded together. The glass filaments are knitted together to form a grid structure having a square aperture of 28 mm and a rib thickness of 4 mm along the MD and CMD. These grids are further bonded mechanically to a non-woven geotextile to form a composite interlayer of about 3 mm thickness as presented in Fig. 2. The composite has an ultimate tensile strength of 28 kN/m (MD) at a strain of 2%, and 25 kN/m (CMD) at a strain of 1.7%.

2.2 Binder Tack Coat and Asphalt Concrete

The binder tack coat used in the current study has a penetration value of 66 and hence, classified as penetration grade (PG) 60/70 bitumen. The various properties of the binder tack coat were determined in the laboratory, and they are as presented in Table 1.

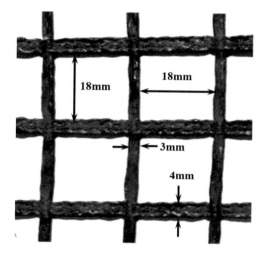

Fig. 1. Polyester grid (PET)

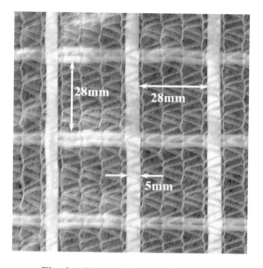

Fig. 2. Glass-grid composite (GGC)

The asphalt concrete mix used in the current study consists of a nominal aggregate size of 13 mm and a PG 60/70 binder. The optimum binder content of the asphalt concrete mix was found to be 5.5% by weight of the aggregates determined from the Marshall Stability test performed as per ASTM D6927. The maximum strength and flow values of asphalt concrete mix prepared at an optimum bitumen content of 5.5% are determined to be 14.25 kN and 2.5 mm, respectively.

Table 1. Properties of binder tack coat

Properties	Values
Penetration (1/10th mm)	66
Specific gravity	1.01
Ductility (cm)	100+
Viscosity, Brookfield at 60 C (centipoise)	460
Softening point (C)	52
Flash point (C)	340
Fire point (C)	365

2.3   Two-Layered Asphalt Specimen Preparation

The two-layered asphalt specimen consists of an old pavement layer, a PG 60/70 binder tack coat, a geosynthetic-interlayer, and an HMA overlay. The old distressed pavement block is extruded from an existing highway during the rehabilitation program and cut into required dimensions of 400 mm length, 300 mm width, and 45 mm thickness. The old pavement block is placed as a bottom layer in the steel mold, a PG 60/70 binder is then applied on the top of old pavement block at a residual rate of 0.25 kg/cm^2. Followed by placement of geosynthetic-interlayers as per the experimental program, a 45 mm thick HMA overlay is compacted with the help of a 5 kg static weight compactor having a height of fall of 500 mm. The unreinforced and geosynthetic-reinforced two-layered asphalt slabs as presented in Fig. 3 are then cut into the beam specimens of 400 mm length, 50 mm width, and 90 mm thickness for the flexural fatigue test. To replicate a crack in the bottom layer, a notch of 40 mm deep (90% of layer thickness) was introduced in the bottom layer of the two-layered asphalt beam specimens.

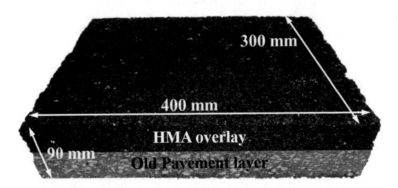

Fig. 3. Two-layered asphalt specimen

Similarly, the two-layered asphalt specimens of 300 mm (length) × 300 mm (width) were prepared as per the procedure explained earlier. A detailed procedure of two-layered asphalt slab preparation for the flexural fatigue and interface shear strength tests are presented by Kumar and Saride (2017), Saride and Kumar (2017), Kumar et al. (2017), and Kumar and Saride (2018b).

3 Experimental Program

The experimental program has been broadly divided into two different stages to evaluate the flexural and shear characteristics of geosynthetic-reinforced asphalt overlays against the unreinforced asphalt overlays. During the first stage, the flexural fatigue characteristics of the unreinforced and geosynthetic-reinforced asphalt overlays placed on a pre-cracked pavement are evaluated with the help of flexural fatigue tests. In the next stage, the shear strength characteristics of the unreinforced and geosynthetic-reinforced asphalt overlays placed on old distressed pavement layer are evaluated using the Interface shear strength tests.

3.1 Flexural Fatigue Test

The flexural fatigue tests were performed under a load-controlled mode, to understand the flexural characteristics of the unreinforced and the geosynthetic-reinforced asphalt overlays. Figure 4 presents the two-layered asphalt beam specimen with a pre-crack of 40 mm depth and 10 mm width, placed in a flexural fatigue test setup used in the study. The load was applied with the help of a computer controlled servo-hydraulic actuator system. A typical haversine load was applied at a frequency of 1 Hz to replicate a live moving traffic condition, corresponding to an equivalent single axle contact pressure of 550 kPa. The load corresponding to a maximum contact pressure of 550 kPa was back-calculated from the maximum flexural stress equation (Eq. 1) derived from the ASTM D7460.

$$\sigma_f = \frac{PL}{bh^2} \tag{1}$$

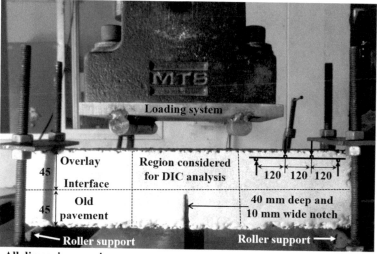

Fig. 4. Flexural fatigue test setup

Where, σ_f is the maximum flexural stress in kPa (550 kPa), P is the maximum load applied in kN, L is the span length of the beam in m, b and h are the width and thickness of the beam in m.

A maximum load of 0.6 kN and a seating load of 0.06 kN was repeatedly applied on the two-layered asphalt specimens until failure, and the mid-span vertical deformations were recorded continuously.

Further, to understand the interface shear strength characteristics of the geosynthetic-reinforced asphalt specimens against the control specimens, the interface shear strength tests were performed.

3.2   Interface Shear Strength Test

The interface shear strength tests are performed on the unreinforced and geosynthetic-reinforced asphalt overlays to understand the interface shear strength characteristics of the asphalt overlays placed on an old pavement layer. The test setup consists of two shear boxes of size 300 mm (length) × 300 mm (width) × 100 mm (depth) separated by an interface zone as presented in Fig. 5. Among the two shear boxes, the bottom shear box is movable, whereas the movement of the top shear box is fixed. A constant normal stress of 30 kPa, 60 kPa, and 120 kPa were applied on the unreinforced and geosynthetic-reinforced specimens, with a constant shear displacement rate of 1 mm/min as per ASTM D5321 and UNI/TS 11214. The tests are performed for all the interface conditions at different normal stress and terminated at the end of a displacement of 50 mm.

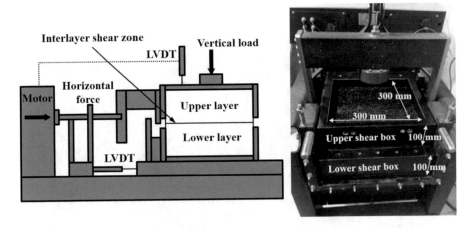

Fig. 5. Interface shear strength test setup

4   Results and Discussion

4.1   Flexural Fatigue Test Results

The flexural fatigue test performed in a load-controlled mode helps to understand the flexural characteristics of the geosynthetic-reinforced asphalt overlays placed on a pre-cracked, old distressed pavement layer. The continuous application of repeated loads on the pre-cracked two-layered asphalt specimens results in an increase in the vertical deformation at the mid-span, with an increase in number of load repetitions (N). The increase in vertical deformation causes the dissipation of energy from the specimen and reduces the specimen stiffness, which further results in complete fracture (failure) of the specimen (Kumar and Saride 2017).

The flexural fatigue test results of a pre-cracked two-layered asphalt beam specimens are presented in Fig. 6, in the form of vertical deformation-number of load cycles plot. It can be observed from Fig. 6, that the control specimens (CS) could not resist a large number of load repetitions before failure (N = 14). Whereas, in contrary,

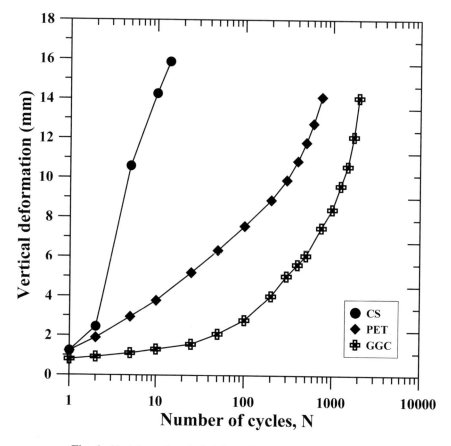

Fig. 6. Variation of vertical deformation with number of cycles

the specimens with geosynthetic-interlayers could resist a large number of load repetitions before failure. For instance, the fatigue life of specimens with PET and GGC interlayers are 750 and 2000 respectively, against a fatigue life of 14 in the control specimens. The change in performances may be attributed to the fact that the presence of geosynthetic-interlayers at the interface of old and new pavement layers restrict the vertical deformation and in turn, restricts the crack growth into the overlays. Among the specimens with geosynthetic-interlayers, the fatigue life of GGC specimen is superior to that of PET specimen. The superior performance of GGC specimens may be attributed to their ability to mobilize an ultimate tensile strength of 28 kN/m at a strain of about 2%. Overall, the geosynthetic-interlayers placed at the interface zone restricts the crack growth into the overlays by absorbing the strain energy (tensile) from the crack tip and dissipates it in the lateral direction at the interface zone. Hence, an improvement in the fatigue life is achieved. It is also to be noted that the dissipation of crack energy in the lateral direction at the interface zone may result in the reduction of interface bond strength and in-turn delaminate the pavement layers.

Further, to quantify the improvement in fatigue life of the geosynthetic-reinforced asphalt specimens against the control specimens, a non-dimensional improvement factor, known as fatigue life improvement ratio (I_{NF}) has been introduced. The fatigue life improvement ratio (I_{NF}) can be defined as a ratio of number of load repetitions sustained by a geosynthetic-reinforced specimen to that sustained by a control specimen, at same vertical deformations, and mathematically expressed as presented in Eq. 2.

$$I_{NF} = \frac{N_R}{N_U} \qquad (2)$$

Where, N_R and N_U are the fatigue lives of geosynthetic-reinforced and unreinforced specimens, respectively

Figure 7 presents the variation of fatigue life improvement ratio with vertical deformation for unreinforced and geosynthetic-reinforced pre-cracked specimens. From Fig. 7, it can be visualized that the I_{NF} increases with an increase in the vertical deformation. The improvement in GGC specimen is observed to be more prominent than that of PET specimens. For instance, in the GGC specimens, the improvement in fatigue life can be witnessed as early as a vertical deformation of 2 mm is reached. Whereas, in specimens with PET interlayer, the improvement ratio is observed to increase after a vertical deformation of 4 mm is reached. These differences in the performances of geosynthetic-interlayers may be attributed to their working tensile properties. The GGC interlayers provide a reinforcing effect at a vertical deformation of 2 mm and continue to provide a reinforcing and stress relieving functions until failure. Besides, the PET interlayer provides the stress relieving effect only after reaching 4 mm vertical deformation and hence, variation in the fatigue life improvements can be witnessed. However, the cracks dissipated in the lateral direction reduces the interface bond strength and may lead to the delamination of pavement layers at the interface zone. Hence, it is also important to understand the interface bond strength characteristics of geosynthetic-reinforced asphalt overlays.

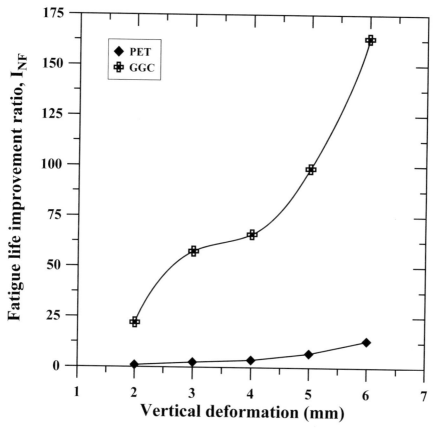

Fig. 7. Variation of improvement in fatigue life with vertical deformation

4.2 Interface Shear Strength Test Results

The interface shear strength tests performed on the all the interface conditions helps to understand their peak and residual interface shear strength characteristics. The interface shear strength test results are presented in the form of shear stress-horizontal displacement output curves as presented in Fig. 8. Figure 8a presents the variation of shear stress with horizontal displacement for different normal stresses of 30 kPa, 60 kPa, and 120 kPa, on an unreinforced interface condition (CS). As expected, the shear stress increases with an increase in the normal stress applied. Similar trends are observed for the specimens reinforced with geosynthetic-interlayers, as shown in Fig. 8b and c for the polyester grid and glass grid composite interlayers, respectively, with a reduction in peak value. It can be observed from Fig. 8 that the shear stress increases with an increase in horizontal displacement until it reaches a peak value and then reduces to a residual state at a horizontal displacement of about 20 mm, irrespective of interface condition and normal stress applied. However, the displacements corresponding to the peak shear stress are observed to vary depending upon the interface conditions, and normal stress applied. For instance, in CS (Fig. 8a), the peak shear stress is achieved at a horizontal displacement of about 4 mm, for the applied normal stress of 30 kPa, 60 kPa, and 120 kPa. Similarly, from Fig. 8c (GGC), it can be observed that with the increase in applied normal stress an increase in the horizontal displacements corresponding to the peak shear stress values are witnessed. Whereas, in the case of PET interface condition (Fig. 8b), there is a decrease in the horizontal displacements corresponding to the peak shear stress value, with the increase in applied normal stress. This condition may be attributed to the increase in the interlocking effect of asphalt overlay, with the apertures of PE interlayer, with an increase in normal stress applied.

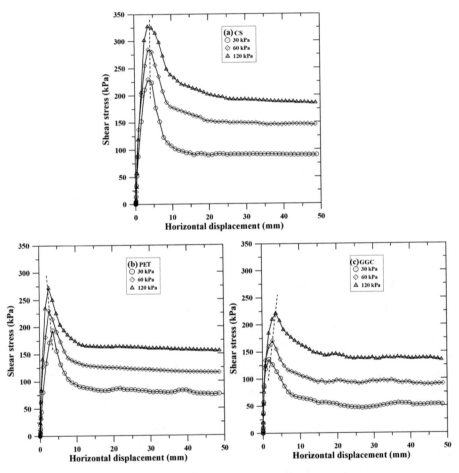

Fig. 8. Variation of shear stress with horizontal displacement: (a) control section (CS); (b) polyester grid coated with polymer modified binder (PET); and (c) glass-grid composite (GGC)

The peak and residual state interface shear strength envelopes obtained for different interface conditions studied in the current research are presented in Fig. 9. It is evident that the interface shear strength between the old and new pavement layers is highest for control interface condition, in both peak and residual states. Among the specimens with geosynthetic-interlayers, the highest interface shear strength is achieved in PET interface condition and finally GGC interface condition, with the least interface shear strength. The reason for this variation in interface shear strength may be due to the presence of apertures in case of PET interlayer. The presence of apertures would help to improve the interface bond condition by a mechanism known as through-hole bonding (THB), which is not witnessed in GGC interlayer, as there are no apertures. The above conditions were witnessed by Ferrotti et al. (2012), Saride and Kumar (2017), and Kumar et al. (2017) in their experimental studies.

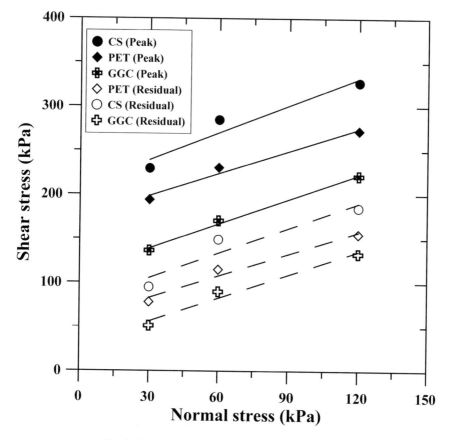

Fig. 9. Variation of shear stress with normal stress

From Fig. 9 and Table 2, it can be clearly distinguished that there is a reduction in the interface shear strength for the interface conditions with geosynthetic-interlayers both at peak and residual states. To evaluate the reduction in interface shear strength, a

Table 2. Summary of interface shear strength test results

Interface condition	Peak state		Residual state	
	c (kPa)	Φ (deg)	c (kPa)	Φ (deg)
Control	208	46	77	43
PET	173	40	58	40
GGC	111	43	30	41

performance factor known as the reduction in interface shear strength (RIS) is introduced. RIS can be mathematically expressed as:

$$RIS(\%) = \frac{IS_C - IS_I}{IS_C} \times 100 \tag{3}$$

Where, IS_C is the interface shear strength between old and new pavement layers for control interface condition, IS_I is the interface shear strength between old and new pavement layers for interlayered interface condition.

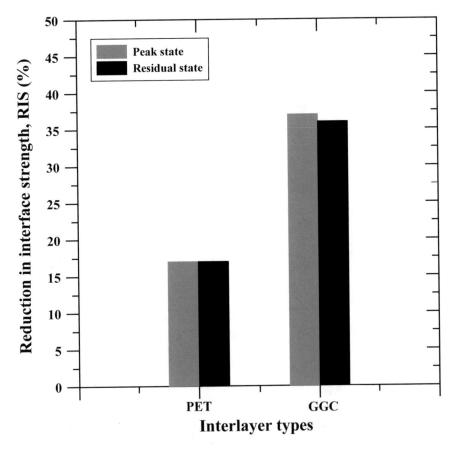

Fig. 10. Variation of RIS with interlayer types

The variation of RIS with different geosynthetic-interlayers for peak and residual states are presented in Fig. 10. From Fig. 10, it can be visualized that GGC interlayered interface condition has the highest reduction in interface shear strength of about 38% and 36% at both peak and residual states, respectively. The reason could be the absence of apertures in GGC interlayer, hence, no direct contact between the pavement layers. The bonding between the old and new pavement layers is now entirely dependent on the adhesion property of the GGC interlayer with the adjacent pavement layers. Besides, a least reduction in interface shear strength of 17% is witnessed in the case of PET interlayered interface condition at both peak and residual states. The polymer modified binder coated on the PET interlayers help to improve the interface shear characteristics by enhancing the adhesion between the interlayer and the adjacent (old and new) layers. Also, the presence of apertures further helps to enhance the interface shear characteristics with the THB mechanism.

Overall, it can be observed that the geosynthetic-interlayers have improved the performance life of asphalt overlays invariably before failure. Among the specimens with geosynthetic-interlayers, the specimens with GGC interlayer performed superior to that of PET interlayered specimens.

5   Conclusions

The flexural and shear characterization of unreinforced and geosynthetic-reinforced asphalt overlays were examined under a two-stage experimental program using flexural fatigue and interface shear strength tests. The following conclusions can be drawn from the study:

- The geosynthetic-interlayers improved the fatigue performance of pre-cracked two-layered asphalt beam specimens. Among them, the performance of GGC specimen is superior to the PET specimen. Owing to the fact that the reinforcing and membrane effects are mobilized completely at a strain value of 2% in GGC interlayers.
- A fatigue life of 750, and 2000 are achieved in PET, and GGC specimens respectively, against a fatigue life of 14 in CS. This accounts for an improvement in fatigue life of about 52-fold and 141-fold in PET and GGC specimens, respectively.
- At the interface zone, the cracks are observed to propagate in the horizontal direction reducing the interface bond strength. The reduction in interface bond strength accelerates the delamination of pavement layers at the interface.
- The interface shear strength test results suggest that the presence of geosynthetic-interlayers at the interface of old and new pavement layers reduce the interface bond strength and may eventually lead to the delamination of pavement layers.
- The interfacial shear strength for the control interface condition is observed to be consistently higher than the geosynthetic-interlayered interface conditions. The reduced interfacial shear strength in the geosynthetic-interlayered specimens may be attributed to the presence of geosynthetic-interlayers, which creates a smooth interface reducing the friction between the pavement layers.

- On an average, a reduction of about 17% and 36% is witnessed in specimens with PET and GGC interlayers, respectively.
- Overall, the geosynthetic-interlayers have improved the fatigue characteristics of the asphalt overlays invariably before failure.

References

Al-Qadi, I.L., Fini, E.H., Elseifi, M.A., Masson, J.F., McGhee, K.M.: Viscosity determination of hot-poured bituminous sealants. Transportation Research Record: The Journal of Transportation Research Board No. 1958, pp. 74–81 (2006)

ASTM D 6927. Standard test method for Marshall stability and flow of asphalt mixtures. Annual Book of ASTM Standards, ASTM International, West Conshohocken, PA

ASTM D4595. Standard test method for determining tensile properties of geotextiles by the wide-width strip method. Annual book of ASTM standards, ASTM International, West Conshohocken, PA

ASTM D5321. Standard test method for determining the coefficient of soil and geosynthetic or geosynthetic and geosynthetic friction by the direct shear method. Annual book of ASTM Standards, ASTM International, West Conshohocken, PA

ASTM D7460. Standard test method for determining fatigue failure of compacted asphalt concrete subjected to repeated flexural bending. Annual book of ASTM Standards, ASTM International, West Conshohocken, PA

Barraza, D.Z., Peréz, M.A.C., Fresno, D.C., Zamanillo, A.V.: Evaluation of anti-reflective cracking systems using geosynthetics in the interlayer zone. Geotext. Geomembr. 29(2), 130–136 (2010)

Caltabiano, M.A.: Reflection cracking in asphalt overlays. Thesis submitted to University of Nottingham for the Degree of Master of Philosophy (1990)

Carey, D.E.: Evaluation of synthetic fabrics for the reduction of reflective cracking. Louisiana Department of Highways, Report No. LA-70-1B (B) (1975)

Cleveland, G.S., Button, J.W., Lytton, R.L.: Geosynthetic in flexible and rigid pavement overlay. Texas Transport Institute, Texas A&M University System, Report No. 1777-1 (2002)

Correia, N.S., Zornberg, J.G.: Strain distribution along geogrid-reinforced asphalt overlays under traffic loading. Geotext. Geomembr. 46(1), 111–120 (2018)

Elseifi, M.A., Al-Qadi, I.L.: A simplified overlay design model against reflective cracking utilizing service life prediction. Road Mater. Pav. Des. 5(2), 169–191 (2004)

Ferrotti, G., Canestrari, F., Pasquini, E., Virgili, A.: Experimental evaluation of the influence of surface coating on fiberglass geogrid performance in asphalt pavements. Geotext. Geomembr. 34, 11–18 (2012)

Gransberg, D.D.: Correlating chip seal performance and construction methods. Transportation Research Record: The Journal of Transportation Research Board No. 1958, pp. 54–58 (2006)

Khodaii, A., Fallah, S., Nejad, F.M.: Effects of geosynthetics on reduction of reflection cracking in asphalt overlay. Geotext. Geomembr. 27(1), 131–140 (2009)

Kim, J., Buttlar, W.G.: Analysis of reflective crack control system involving reinforcing grid over base-isolating interlayer mixture. J. Transp. Eng. ASCE 128(4), 375–385 (2002)

Kumar, V.V.: Behavior of geosynthetic-interlayered asphalt overlays. Ph.D. thesis submitted to Indian Institute of Technology Hyderabad, India (2017)

Kumar, V.V., Saride, S.: Use of digital image correlation for the evaluation of flexural fatigue behavior of asphalt beams with geosynthetic interlayers. Transportation Research Record: The Journal of Transportation Research Board No. 2631, pp. 55–64 (2017)

Kumar, V.V., Saride, S.: Influence of crack-depth on the performance of geosynthetic-reinforced asphalt overlays. In: Proceedings on International Symposium of Geotechnics in Transport Infrastructure (ISGTI)-2018, IIT Delhi, India (2018a)

Kumar, V.V., Saride, S.: Evaluation of cracking resistance potential of geosynthetic reinforced asphalt overlays using direct tensile strength tests. Constr. Build. Mater. **162**(20), 37–47 (2018)

Kumar, V.V., Saride, S., Peddinti, P.R.T.: Interfacial shear properties of geosynthetic interlayered asphalt overlays. In: Proceedings on Geotechnical Frontiers-2017, Orlando, USA (2017)

Prieto, J.N., Gallego, J., Perez, I.: Application of the wheel reflective cracking test for assessing geosynthetics in anti-reflection pavement cracking systems. Geosynth. Int. **14**(5), 287–297 (2007)

Sanders, P.J., Brown, S.F., Thom, N.H.: Reinforced asphalt for crack and rut control. In: Seventh Conference on Asphalt Pavements, Southern Africa (1999)

Saride, S., Kumar, V.V.: Influence of geosynthetic-interlayers on the performance of asphalt overlays placed on pre-cracked pavements. Geotext. Geomembr. **45**(3), 184–196 (2017)

Sherman, G.: Minimizing reflection cracking of pavement overlays. Technical report, NCHRP Synthesis 92 (1982)

UNI/TS 11214. Mechanical properties of road and airfield pavements: interlayer shear performance-related characterization: ASTRA Test Method

Virgili, A., Canestrari, F., Grilli, A., Santagata, F.A.: Repeated load test on bituminous systems reinforced by geosynthetics. Geotext. Geomembr. **27**(3), 187–195 (2009)

Walubita, L.F., Faruk, A.N.M., Zhang, J., Hu, X.: Characterizing the cracking and fracture properties of geosynthetic interlayer reinforced HMA samples using the Overlay Tester (OT). Constr. Build. Mater. **93**(15), 695–702 (2015)

Wargo, A., Safavizadeh, S.A., Kim, Y.R.: Comparing the performance of fiberglass grid with composite interlayer systems in asphalt concrete. Transportation Research Record: The Journal of Transportation Research Board No. 2631, pp. 123–132 (2017)

Evaluation of Strength and Resilient Modulus Characteristics of Fly Ash Geopolymer Stabilized Reclaimed Asphalt Pavement Material

Maheshbabu Jallu[✉] and Sireesh Saride

Department of Civil Engineering, Indian Institute of Technology
Hyderabad (IITH), Kandi, Telangana, India
{ce15resch11014, sireesh}@iith.ac.in

Abstract. Utilization of sustainable road construction materials has been the focus of research worldwide in recent times. Virgin aggregate is a primary material in the pavement industry; hence, finding an alternative is of extreme importance, concerned with the more prudent use of natural resources and the protection of the environment. The present research explored the usage of a significant portion of reclaimed asphalt pavement (RAP), activated with low calcium fly ash (FA) as a binding material. A liquid alkaline activator comprising sodium silicate solution (Na_2SiO_3) and sodium hydroxide (NaOH) was used for the alkali activation of the mix. The fundamental design parameters including Unconfined Compressive Strength (UCS) and resilient modulus (M_r) characteristics of the stabilized RAP:VA+FA geopolymer specimens were studied at room temperature. The resilient modulus (M_r) value in mechanistic-empirical analyses has been widely accepted in design/analysis of the pavement structures. Therefore, the present study aims to examine the resilient behaviour of the pavement base material stabilized with alkali activated low calcium Indian fly ashes, obtained from the southern region of India. The effect of additives on the microstructure of RAP:VA+FA blends were verified for one day and 28 days cured samples using X-ray diffraction (XRD) studies. Since the UCS and M_r values met the specified strength requirements, the stabilized mix can be used as a pavement base material.

1 Introduction

In recent times, increasing attention has been drawn towards the utilization of sustainable construction materials in India and around the world. The potential application and wide usage of these materials in the construction industry will lead to the significant environmental and economic benefits. Reclaimed Asphalt Pavement (RAP) is one of the sustainable alternatives to the virgin aggregate in the pavement applications. The utilization of RAP has grown extensively, lessening the use of virgin aggregate concerned with a sensible use of natural resources. Two common type of recycling methods are available for the removal of aged/distressed pavements are cold and hot milling. Among these, the cold milling process is the most preferred one. Properly

© Springer Nature Switzerland AG 2019
M. Meguid et al. (Eds.): GeoMEast 2018, SUCI, pp. 124–136, 2019.
https://doi.org/10.1007/978-3-030-01944-0_10

milled and screened RAP consists of high-quality, well-graded aggregate and it can be widely utilized in pavement base subbase applications (Arulrajah et al. 2013, 2014; Saride et al. 2014). However, a large quantity of RAP not being used in any of these applications due to its inferior strength and stiffness characteristics (Hoyos et al. 2011). Hence, before any application, the usage of RAP requires a thorough understanding of its physical and mechanical strength characteristics. The usage of RAP in base/subbase applications often does not meet the local authority requirements although it is blended with the other high-quality natural aggregate. Therefore, the strength/stiffness characteristics can be improved by stabilizing it with a binding material (Taha et al. 2002; Puppala et al. 2011; Saride et al. 2014). Taha et al. (2002) studied the cement stabilized RAP and RAP-virgin aggregate blend as a pavement base material and concluded that a 100% RAP could not be utilized as a pavement base material unless it is stabilized with cement. Puppala et al. (2011) conducted a series of resilient modulus tests (M_r) on cement stabilized RAP specimens and found that the RAP stabilized with cement could improve the performance the pavement base layer. Suebsuk et al. (2017) investigated the UCS of the RAP-marginal lateritic soil (LS) blend stabilized with cement and reported that the improved strength had met the requirements of the department of highways (DOH), Thailand. However, given the sustainability and environmental ethics, the cement treated RAP is not considered as an eco-friendly/sustainable alternative. It is also important to note that, the cement industry alone generates 5–7% of CO_2 emissions globally (Humphreys and Mahasenan 2002). Hence, there is a pressing need to utilize the other sustainable alternatives to stabilize the secondary aggregate material.

Fly ash has been recognized as an alternative sustainable binding material and abundantly available by-product from the coal-based thermal power plants. It primarily contains silica, alumina with little or no pozzolanic properties (Lee and Van Deventer 2002). Several researchers have stated that inactive pozzolans can be activated in an alkaline environment to form the pozzolanic compounds. Initially, Palomo et al. (1999) described the mechanism of the alkali-activated fly ash with alkaline solutions such as NaOH, KOH. It was concluded that the curing time and temperature together with the solution/fly ash ratio are the critical variables in the development of mechanical strength of the final product. Puertas et al. (2000) studied on fly ash/slag mortar activated with various NaOH molar concentrations from 2 M and 10 M and concluded that the strength development is directly related to the NaOH concentration, and higher compressive strengths were achieved when the NaOH molar concentration was 10 M. Saride et al. (2014) used high calcium fly ash obtained from the Neyveli thermal power station to stabilize the RAP-virgin aggregate mixture. In this study, 10, 20, 30 and 40% dosages of fly ash were used and concluded that 80RAP:20VA+40FA has met both the UCS and Mr requirements of base/subbase material for low volume roads. Avirneni et al. (2016) used NaOH activated fly ash to treat the RAP:VA blend for pavement base/subbase applications. This study confirms that the usage of 0.5 M and 1 M NaOH solution could meet the strength requirements as specified by the IRC: 37-2012 for high-volume roads. Jimenez and Palomo (2005) and Criado et al. (2007) stated that, a mixture of sodium silicate and a specified molar concentration of NaOH solution could form the potential alkaline environment in the source material. Thus it leaches the silica and alumina resulting the Sodium aluminosilicate hydrate (N-A-S-H) gel in the system. Hoy et al. (2016a) studied on strength and microstructural characteristics of RAP stabilized with high calcium fly ash for pavement base applications. In this study, the design mix was activated with liquid alkaline activator (LAA) solution comprises the mixture of NaOH (10 M): Na_2SiO_3 and concluded that the LAA ratio of $50NaOH:50Na_2SiO_3$ produced the better UC strength and observed a dense microstructure in the scanning electron microscopy (SEM) analysis. Phummiphan et al. (2015) found the maximum UCS at LAA ($Na_2SiO_3:NaOH$ (5 M)) ratio of 50:50 to stabilize the marginal lateritic soil (LS) with high calcium fly ash. Hoy et al. (2016b) studied on strength and toxic leaching of the alkali-activated RAP:FA blend and reported that the maximum UC strength was found at $50Na_2SiO_3:50NaOH$ (5 M)) activator combination. Saride et al. (2016) conducted a series of resilient modulus tests on high calcium fly ash stabilized RAP:VA specimens. They have tested on various mixtures viz. 100RAP:0VA, 80RAP:20VA and 60RAP:40VA stabilized with 10, 20, 30 and 40% dosage of fly ash. Results indicated that 60RAP:40VA+40FA mix combination has shown the better results. Also, they found an increase in Mr upon increasing the FA dosage to 40%. Kua et al. (2017) studied the resilient modulus characteristics of geopolymer stabilized spent coffee grounds (CG). It is observed that the geopolymerization drastically increased the Mr of the CG and this green material can be used in the pavement subgrades.

However, from the above studies, it is unclear that the stiffness characteristics of the fly ash geopolymer stabilized RAP as a pavement base material. Majority of studies were considered the UCS as an indicator to design the pavement base course. It is also important to note that the UCS based design may not perform as a pavement layer. The layer deformations are recoverable after the post-compaction. Therefore, it is vitally important to study the resilient modulus characteristics of the base material as specified by the AASHTO T 307 (2003). Also, in the context of RAP systems, the basis of selection of the NaOH molarity in the working alkali activator solution is unclear.

2   Objectives

Limited studies are available on the usage of alkali-activated fly ash (AAF) stabilized RAP as a pavement base material. The main objective of the present study is to verify the UCS and Mr characteristics of various design mixes as specified in IRC 37: 2012 for pavement base. This study also verifies the optimum NaOH molar concentration required to maintain a potential alkaline environment (pH $\sim$ 12.6). Based on available literature, a mix proportion of 60:40 (RAP:VA)+20%FA, represented as 60R:40V +20FA was selected in this study. Further, the XRD studies were conducted to identify the mineral microstructure of the inactive fly ash and the design mixtures.

3 Materials and Methods

3.1 Materials

In this study, RAP has been collected from an ongoing cold milling operation near Rajahmundry city, Andhra Pradesh, India. The collected RAP was brought to the IIT Hyderabad campus, and sieved into different sizes then placed into separate bins. As per the USCS classification, current RAP material comes under well-graded gravel (GW), and it is shown in the Fig. 1. The virgin aggregate material has been gathered from the quarry near Sangareddy, Telangana. Two different fly ashes used in this study were collected from Vijayawada (FA-V), and Ramagundam (FA-R) thermal power plants in the south India. As per the ASTM C618, these two fly ashes can be classified as low calcium ($SiO_2 + Al_2O_3 + Fe_2O_3 > 50\%$ and $CaO < 10\%$, Class-F) fly ash. The oxide compositions of these fly ashes were determined by using X-ray fluorescence spectroscopy (XRF) are listed in Table 1, and the particle size distribution curves are shown in the Fig. 1.

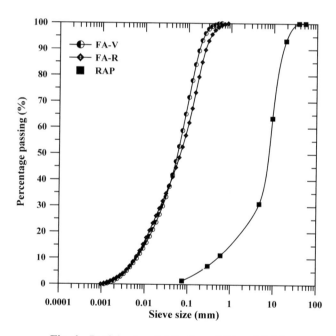

Fig. 1. Particle size distribution of FA and RAP

3.2 Preparation of Liquid Alkaline Activator (LAA)

The alkaline activator being used to initiate the pozzolanic reactions among the active oxide compositions of fly ash. A liquid alkaline activator (LAA) consists of a mixture of sodium silicate (Na_2SiO_3) solution and specified molarity of sodium hydroxide (NaOH). Commercial grade sodium hydroxide flakes with 98% purity were used to

Table 1. Oxide Compositions (% by Mass) of fly ashes

Oxide composition (%)	Vijayawada	Ramagundam
SiO_2	55.38	53.79
Al_2O_3	29.21	26.82
Fe_2O_3	5.83	5.4
CaO	2.05	5.8
TiO_2	2.5	2.26
K_2O	1.67	2.13
MgO	1.47	1.68
P_2O_5	0.73	0.82
SO_3	0.37	0.66

prepare the desired molar concentration was obtained from the Madhava laboratories, Hyderabad. Sodium silicate solution with the chemical composition of 10% Na_2O, 38% SiO_2 and 52% H_2O was used to prepare the liquid alkaline solution, and it was obtained from Sarada silicates, Sangareddy, Hyderabad. The specific gravity of sodium silicate solution is 1.5 g/cm^3. The effect of different activator ratios (Na_2SiO_3:NaOH) on 60R:40V+20FA mixtures were studied in terms of UCS, Mr and microstructural characteristics

3.3 Mixture Proportions

Based on the available literature (Kua et al. 2016; Hoy et al. 2016a; Phummiphan et al. 2016) and the preliminary experiments, the following LAA combinations and the molarity ranges were selected to prepare the mix design.

- Alkali hydroxide i.e. NaOH and sodium silicate solution.
- The ratio of sodium silicate to sodium hydroxide (Na_2SiO_3:NaOH) = 30:70, 50:50, 70:30 and 90:10.
- Molarity of NaOH = 0.5 M and 3 M.
- Aggregates constituted RAP and virgin aggregate.

3.4 Preparation of Design Mix for UCS and Mr Testing

Dry aggregate (RAP+VA) and fly ash additives were selected as per the required weights. The preparation of LAA solution releases heat since the reaction was exothermic. Therefore, before mixing it to the dry mixture, it is left at room temperature for about 24 h to bring down to the room temperature. The dry mix was obtained by mixing RAP, virgin aggregate and fly ash in 60R:40V+20FA combination. After the thorough mixing of aggregates with LAA, the specimens were casted in 100 mm diameter and 200 mm height cylindrical moulds. Later, the samples were allowed to be in the moulded state and placed in a humidity chamber for 24 h. The moulded specimens were then de-moulded and wrapped with an aluminium foil to arrest the moisture evaporation and then placed in a humidity chamber for further curing.

4   Experimental Results

4.1   Compaction Testing

The modified Proctor compaction tests in accordance with the ASTM-D 1557-12 were carried out. Figure 2 indicates the relationship between dry unit weight and moisture content of the compacted RAP:VA+FA blends. From these two mixes, it is evident that the compaction curve related to the Ramagundam fly ash based mix has shifted towards the upper side which indicates an increase in the dry density. This is may be due to the finer particles have more specific surface area, and thus it occupies the maximum void space in the mixture.

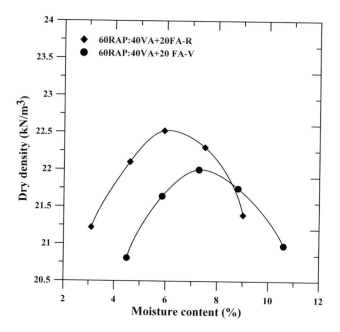

Fig. 2. Compaction curves or 60R:40V+20F mixes with different fly ashes

4.2   pH Studies

Since high pH concentration favors the pozzolanic reactions, the pH of the design mix plays a vital role in the geopolymerization. In general, at high alkaline environment, the OH^- ions attributed to the dissolution of inactive pozzolans present in the source material (Bell 1996). As the NaOH molarity increases, the pH of the mix rises due to the addition of OH^- ions. Therefore, the dissolution of excess Si^{4+}, Al^{3+}, and Ca^{2+} ions will take place in the mix leading to form the calcium-aluminum-silicate-hydrate (C-A-S-H)/sodium aluminosilicate hydrate (N-A-S-H) gel in the system.

In this study, the pH of the powdered sample comprises the mixture of RAP, virgin aggregate and the fly ash mixture treated with various molar concentrations of NaOH was studied. The pH of the mix determined in accordance with the ASTM D 4972 (2013). Figure 3 details the variation of pH with the design mixtures prepared with different NaOH molar concentrations. The NaOH molarity ranging from 0.5 M to 6 M was studied. From this figure, it is evident that the pH of the mixtures without activation is 8 and 9.3. Mixtures prepared with the Ramagundam fly ash with 0.5 molar NaOH exhibits a pH around 12.3 whereas it is 11.5 in the Vijayawada fly ash based mix. Further increment in the of NaOH molarity (at 3 M) of the blend shows a pH of 12.6 and 12.75, and this is more than the desired pH to maintain the potential alkaline environment (Eades and Grim 1999). Based on this, a set of initial UCS studies were conducted and found that 0.5 M and 3 M NaOH concentrations were optimum for Ramagundam, and Vijayawada fly ash based mixes. Therefore, these molar concentrations were adopted while preparing the UCS and M_r test samples.

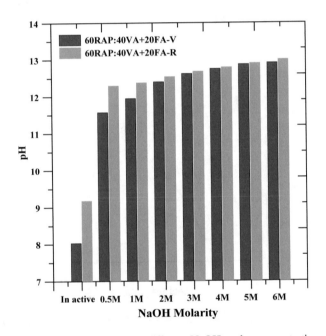

Fig. 3. Variation of pH with different NaOH molar concentrations

4.3   Unconfined Compressive Strength (UCS) Tests

The UCS specimens were prepared in accordance with the ASTM D1557 and placed in a humidity chamber under controlled moisture (95% Relative Humidity) and room temperature (27 °C). According to IRC:37-2012 (IRC 2012), the UC strength of 28 day's specimen to be used in high volume roads should be minimum 4.5 MPa. A series of UCS tests were conducted on each fly ash design mix on different Na_2SiO_3: NaOH ratios (30:70 to 90:10). The design mixtures were prepared using Vijayawada and Ramagundam fly ashes, and the corresponding 28 days UC strength data is presented in the Fig. 4. From this figure, it can be noticed that the UCS of both the fly ashes based

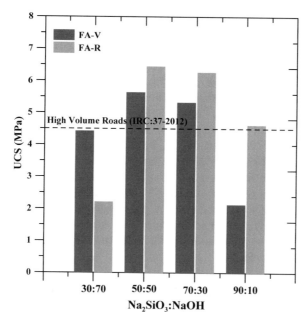

Fig. 4. Variation of 28 days UCS for different LAA ratio with FA-V and R

mixtures was increased with curing time. However, it shows slightly lower values in the Vijayawada fly ash based mixtures.

This decreasing trend in Vijayawada fly ash can be attributed to the lesser reactive potential as evidenced by Singh et al. (2018). The RAP:VA+FA based geopolymer specimens synthesized with the Ramagundam fly ash based mixtures exhibited relatively higher compressive strengths with a maximum UCS of 6.5 MPa. This was expected because of the considerable reactive portion, and active minerals are presented in the raw fly ash, where the Si^{4+}, $Al^{3+,}$ and Ca^{+2} ions are dissolved from the surface of the particles in contact with the alkaline solution. From the XRF test results, it is clear that the CaO content in Ramagundam fly ash is (5.8%) reasonably good along with the alumina and silica. It is also important to note that the UC strengths in both the fly ash based specimens were attained maximum strength at $50Na_2SiO_3:50NaOH$ and decreased later (in $70Na_2SiO_3:30NaOH$ and $90\ Na_2SiO_3:10NaOH$ combinations). This decreasing in the UC strength may be due to the supplement of additional sodium silicate than the optimum. Therefore, it arrests the moisture evaporation and hinders the structure formation (Cheng and Chiu 2003).

4.4 Resilient Modulus (Mr) Tests

A series of resilient modulus (Mr) tests were performed using the cyclic triaxial on the RAP:VA+FA specimens by applying a sequence of cyclic loading as specified by the AASHTO T 307-99. Samples prepared with 100 mm × 200 mm dimensions were used to test the Mr characteristics. A typical cyclic triaxial test setup has been shown in the Fig. 5, which contains the 50 kN load cell, hydraulic pressure unit (HPU) and the air pressure controller (APC) devices. A haversine-shaped wave load with a loading period of 0.1 s and a resting period of 0.9 s was applied. Tests were performed to assess the resilient modulus of specimens prepared with Vijayawada and Ramagundam fly ashes with 60RAP:40VA+20FA combination as similar to the UCS design mix. The alkaline activator ratio (Na_2SiO_3:NaOH) was varied from $30Na_2SiO_3$:70NaOH to $90Na_2SiO_3$:10NaOH. The NaOH molar concentration in each LAA ratio was considered as 0.5 M, and 1 M for Ramagundam and Vijayawada fly ashes respectively. As per the IRC:37-2012, the Mr of the 28 days' specimen's ≥ 450 MPa is desired to qualify as a cemented base material in the pavement base layer. In each test, the sample was subjected to the to five different confining pressures and three different deviatoric stresses per confining pressure as stated in the AASHTO test procedure for base materials. The effect of source of fly ash and the amount alkaline activator on the Mr of RAP:VA specimens plays an important role. As observed in the UCS test results, the source of fly ash has the remarkable effects on its UC strength and also the Mr values has shown the similar effect.

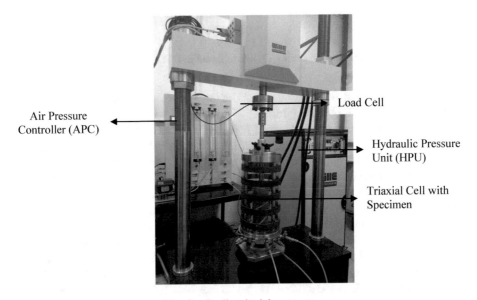

Fig. 5. Cyclic triaxial test setup

From the Fig. 6, it can be noticed that the Mr values of $90Na_2SiO_3$:10NaOH combination are relatively low. This is due to the presence of higher amount of sodium silicate solution (Na_2SiO_3) presented in the 90:10 lead to the lower stiffness. Specimens prepared with the Vijayawada fly ash shows relatively lesser Mr values than the Ramagundam fly ash based samples. This can be due to the higher amount of inactive oxide compounds presented in the Vijayawada fly ash. In general, the reactive oxide compounds such as silica and alumina in the alkali-activated fly ash will participate in the geopolymerization. Thus it forms the N-A-S-H/C-A-S-H gels in the system.

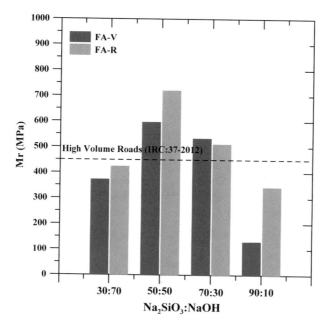

Fig. 6. Variation of 28 days Mr for different LAA ratio with FA-V and R

Therefore, during the hydration process, the formation of geopolymer gels in the system leads to the higher Mr values in the Ramagundam fly ash based mixture.

4.5 XRD Analysis

X-Ray Diffraction (XRD) analysis has been performed to identify the available minerals in the original fly ash (fly ash-V), one day and 28 days crushed sample of 60R:40V+20FA-V mixture activated with $50Na_2SiO_3:50NaOH$ (3 M). The XRD spectrums were obtained using $CuK\alpha$ radiation on a D2 phaser (Bruker) diffractometer, operated at ten mA and 30 kV. The regular XRD scans are $(\theta-\theta)$ geometry range of $10°$ and $70°$ at $0.03°$ 2θ step size per second. Therefore, each scan took around 34 min and to minimize the errors; two tests were performed per sample. Figure 7 represents the effective mineral compounds in this fly ash viz. Quartz, Mullite, Hematite, and Magnetite. Quartz (SiO_2) and Mullite ($3Al_2O_3.\ 2SiO_2$) are the primary mineral compounds to form N-A-S-H gel during the geopolymerization process. From this figure, it can be observed that the peak intensities were significantly increased during the hydration process and new mineral compounds were detected viz. Albite and Dolomite in the region of $15°$ to $40°$. The Albite mineral ($NaAlSi_3O_8$) contains the sodium (Na), aluminium (Al), silica (Si) and oxygen compounds. It implies that the formation of N-A-S-H phases as a result of the alkaline activation. Dolomite [$CaMg(CO)_2$] majorly consists of the magnesium and calcium cementitious compounds which result in the strength enhancement.

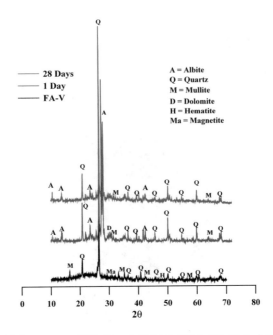

Fig. 7. XRD patterns of FA-V and 60RAP:40VA+FA-V design mixtures

5   Conclusions

In this paper, a series of experiments were conducted to examine the strength, stiffness and microstructural characteristics of 60RAP:40VA+20FA mixes. The effects source of fly ash, the ratio of Na₂SiO₃:NaOH and the NaOH molar concentration into various fly ash design mixes were analysed. The extent of formation of new mineral structures were verified with X-ray diffraction (XRD), and the conclusions are as follows:

The pH studies show the adequate amount of NaOH (in molars) to obtain the desired UCS as per the IRC:37-2012 for the Indian high volume roads. The effect of NaOH molar concentration on the mechanical properties was significant due to the geopolymerization.

The addition of LAA into the dry mix has significantly improved the UC strength as it contains a decent amount of reactive silica. The optimum LAA plays a major role in the strength gain. Thus the 50Na₂SiO₃:50NaOH combination showed the better UCS and Mr values.

The XRD spectrum for the 28 days hydrated sample showed a clean amorphous hump in Vijayawada fly ash based design mix, whereas it showed sharp crystalline humps for the un-hydrated mix and the original fly ash.

Based on the laboratory and field studies, it can be concluded that the fly ash stabilized RAP can be utilized as a sustainable pavement base/subbase material in the Indian high volume roads.

References

AASHTO: Standard Method of Test for Determining the Resilient Modulus of Soils and Aggregate Materials. T-307-99, Washington, D.C. (2003)

Arulrajah, A., et al.: Geotechnical and geoenvironmental properties of recycled construction and demolition materials in pavement subbase applications. J. Mater. Civ. Eng. **25**(8), 1077–1088 (2013)

Arulrajah, A., et al.: Reclaimed asphalt pavement and recycled concrete aggregate blends in pavement subbases: laboratory and field evaluation. J. Mater. Civ. Eng. **26**(2), 349–357 (2014)

ASTM D4972-13: Standard Test Method for pH of Soils. ASTM International, West Conshocken, PA (2013)

Avirneni, D., et al.: Durability and long-term performance of geopolymer stabilized reclaimed asphalt pavement base courses. Constr. Built. Mater. **121**, 198–209 (2016)

Bell, F.G.: Lime stabilization of clay minerals and soils. Eng. Geol. **42**, 223–237 (1996)

Criado, M., et al.: An XRD study of the effect of the SiO_2/Na_2O ratio on the alkali activation of fly ash. Cem. Concr. Res. **37**(5), 671–679 (2007)

Eades, J.L., Grim, R.E.: A quick test to determine lime requirements for lime stabilization. Highway Research Record No. 3. National Academy of Sciences. Highway Research Board, Washington, D.C. (1999)

Fernández-Jiménez, A., Palomo, A.: Composition and microstructure of alkali activated fly ash binder: effect of the activator. Cem. Concr. Res. **35**, 1984–1992 (2005)

Hoy, M.L., et al.: Strength development of recycled asphalt pavement-fly ash geopolymer as a road construction material. Constr. Built. Mater. **117**, 209–219 (2016a)

Hoy, M.L., et al.: Recycled asphalt pavement–fly ash geopolymers as a sustainable pavement base material: Strength and toxic leaching investigations. Sci. Total Environ. **573**, 19–26 (2016b)

Hoyos, L.R., et al.: Characterization of cement-fiber-treated reclaimed asphalt pavement aggregates: preliminary investigation. J. Mater. Civ. Eng. **23**, 977–989 (2011)

Humphreys, K., Mahasenan, M.: Toward a Sustainable Cement Industry. Substudy 8: Climate Change. World Business Council for Sustainable Development, Geneva (2002)

Indian Roads Congress (IRC): Guidelines for the Design of Flexible Pavements. Indian Code of Practice, IRC, 37 (2012)

Kua, T.A., Arulrajah, A., Horpibulsuk, S., Du, Y.J., Shen, S.L.: Strength assessment of spent coffee grounds-geopolymer cement utilizing slag and fly ash precursors. Constr. Build. Mate. **115**, 565–575 (2016)

Kua, T.A., et al.: Stiffness and deformation properties of spent coffee grounds based geopolymers. Constr. Build. Mater. **138**, 79–87 (2017)

Lee, W.K.W., Van Deventer, J.S.J.: Structural reorganisation of class F fly ash in alkaline silicate solutions. Colloids Surf. A **211**(1), 49–66 (2002)

Palomo, A., et al.: Chemical stability of cementitious materials based on metakaolin. Cem. Concr. Res. **29**(7), 997–1004 (1999)

Phummiphan, I., Horpibulsuk, S., Sukmak, P., Chinkulkijniwat, A., Arulrajah, A., Shen, S. L.: Stabilisation of marginal lateriticsoil using high calcium fly ash based geopolymer. Road Mater. Pave-ment Des. **17**(4), 877–891 (2015). https://doi.org/10.1080/14680629.2015.1132632

Phummiphan, I., et al.: Marginal lateritic soil stabilized with calcium carbide residue and fly ash geopolymers as a sustainable pavement base material. J. Mater. Civ. Eng. **29**(2), 04016195 (2016)

Puertas, F., et al.: Alkali-activated fly ash/slag cement strength behaviour and hydration products. Cem. Concr. Res. **30**, 1625–1632 (2000)

Puppala, A.J., et al.: Resilient moduli response of moderately cement-treated reclaimed asphalt pavement aggregates. J. Mater. Civ. Eng. **23**(7), 990–998 (2011)

Saride, S., et al.: Evaluation of fly ash treated reclaimed asphalt pavement for base/subbase applications. Indian Geotech. J. **45**, 401 (2014). https://doi.org/10.1007/s40098-014-0137-z

Saride, S., et al.: Micro-mechanical interaction of activated fly ash mortar and reclaimed asphalt pavement materials. Constr. Built. Mater. **123**(2016), 424–435 (2016)

Suebsuk, J., et al.: Strength prediction of cement-stabilized reclaimed asphalt pavement and lateritic soil blends. Int. J. Pavement Eng. (2017). https://doi.org/10.1080/10298436.2017.1293265

Taha, R., et al.: Cement stabilization of reclaimed asphalt pavement aggregate for road bases and subbases. J. Mater. Civ. Eng. ASCE **14**(3), 239–245 (2002)

How Stiffness of Reinforcement Affects the Type of Major Reinforcement Force Developed at Various Orientations in Reinforced Sand?

B. A. Mir[✉] and R. Shah

Department of Civil Engineering, National Institute of Technology Srinagar,
Srinagar, Jammu and Kashmir 190006, India
p7mir@nitsri.net, rufaidahshah@gmail.com

Abstract. Reinforced soil is any soil system in which some reinforcing elements called inclusions are placed to improve its mechanical properties. The reinforcement acts by restraining the deformation of the soil mass achieved by soil-reinforcement interface bonding. It results in increased stiffness and consequently the strength by increasing confinement and reducing dilation tendency of the soil mass. Large scale experimental study on soil-reinforcement interaction in the literature supported the conclusion that the improvement in soil strength attained from the bending stiffness of the reinforcement is always small as compared to the improvement obtained from the axial capacity of reinforcement. The objective of the current investigation is to evaluate the effect of relative stiffness of the reinforcement on the type of major reinforcement force developed in the reinforcement at various orientations of reinforcement with respect to shear plane using small-scale direct shear tests. In this study, a series of small size direct shear tests were conducted on clean, poorly graded medium grained sand under undrained condition. Two types of reinforcements of annealed rigid reinforcement binding wire and flexible Geogrid (SGi-100) strips were used of relatively varying stiffnesses. The reinforcements were only embedded in the sand and were not anchored to the shear box at any point. The sand was tested in loose and medium dense states with reinforcement oriented at various angles with the vertical plane. Based on the test results, it was concluded that the attainment of maximum increase in shear strength was observed for specific (optimum) orientations with the vertical for both inward and outward orientations and that the stiffness of reinforcement influenced the optimum orientation of reinforcement corresponding to the attainment of maximum shear strength in terms of both, the composite friction angle as well as the apparent cohesion developed due to reinforcement in the reinforced soil mass.

Keywords: Cohesionless soils · Reinforced soil
Soil-reinforcement interaction · Reinforcement stiffness
Reinforcement orientation · Direct shear test

© Springer Nature Switzerland AG 2019
M. Meguid et al. (Eds.): GeoMEast 2018, SUCI, pp. 137–151, 2019.
https://doi.org/10.1007/978-3-030-01944-0_11

1 Introduction

Unlike compression behavior of soils, the tensile behavior plays a significantly important role in various engineering applications. However, soils are weak in tension. Thus, compacted soils used in various construction activities can be subjected to tensile failure. Therefore, to increase the tensile behavior of soils, some reinforcing elements called inclusions are used to increase the strength and stiffness of the soil to improve the mechanical properties of the reinforced soil, also known as composite soil. The reinforcement acts by restraining the deformation of the soil mass achieved by soil-reinforcement interface bonding.

For the purpose of soil reinforcement, there has been a rapid development in the use of goesynthetics. They possess high tensile strength and are easy to install having rapid installation procedure with low installation cost. Since these materials do not possess much flexural stiffness and act as flexible reinforcements, the increase in shear strength due to their presence in the soil mass can be attributed to the tensile strains that are mobilized in the reinforcement. The adhesion between soil and a geosynthetic material and the friction angle of the interface are important parameters in the analysis and design of such structures (Hsieh et al. 2011). There is a positive correlation between the shear strength ratio, τ/σ and the tensile strength of the geogrid (Lui et al. 2009). So, one can expect maximum improvement when the orientation of reinforcement coincides with the direction of tensile strains developed in the unreinforced soil. However, the use of rigid reinforcement having high flexural strength (such as metallic strips, soil nails, etc.) as reinforcing elements have attracted a much debate over the type of major reinforcement force developed in these elements. In addition to the axial stresses developed in the reinforcing element, the increase in shear strength can be attributed to the bending moments developed in it due to its flexural stiffness. Direct shear tests have been the most suitable to study the soil-reinforcement interactions (Horpibulsuk and Niramitkornburee 2010) because they can more accurately simulate the mechanism of shear that takes place along a potential failure surface in a reinforced earth structure as shown in Fig. 1.

Jewell (1980) was the first to use the direct shear apparatus for studying soil-reinforcement interaction and performed a series of tests using grid and bar reinforcements. The work was focused on studying the effects of flexible tensile reinforcement on the mechanical behavior of sand. It has been concluded that the maximum efficiency could be achieved for a relatively flexible tensile reinforcement when the orientation was approximately along the direction of principle tensile strain in the unreinforced soil. The reinforcement acts by restraining the tensile strains developed in the soil mass. McGown et al. (1978) obtained a similar result using plane strain cell test on sand reinforced with a single layer of flexible reinforcement. Dyer in 1985 used a photo elastic technique to investigate the stresses developed between the reinforcement and an artificial granular material (crushed glass) in the direct shear test. The findings of Dyer were in agreement with the conclusions drawn by McGown et al. (1978) and Jewell (1980). As such it was established that dilatancy of soil and reinforcement extensibility significantly affects the tensile forces generated in the inclusions (Dyer 1985; McGown et al. 1978). Assuming the behavior of soil like that of an

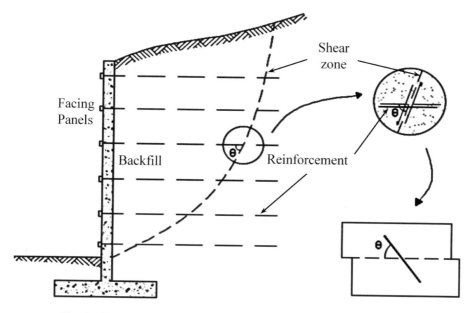

Fig. 1. Similarity between reinforced earth structure and direct shear test

elasto-plastic strain hardening material while that of the reinforcement as elastic-perfectly plastic material, Juran et al. (1988) proposed a load transfer model to integrate the effects of soil dilatancy and reinforcement extensibility to analyse the response of sand reinforced with tensile reinforcements and apply it for numerical simulation of direct shear tests on reinforced sand. The model was later extended for cohesive soils and its validity was tested using modified direct shear test for residual soils. Predictions using the model were in agreement with the experimental results that confirmed the strain hardening behavior of reinforced soil particularly for high stress levels. (Mofiz et al. 2004).

Yang (1972) and Schlosser and Long (1974) investigated the mechanics of reinforced mass using the concept of apparent cohesion and apparent confining pressure, and reported that apparent cohesion can be determined using strength data for unreinforced soil by means of concept of apparent confining pressure. Further, using this concept, Bauer and Zhao (1993) performed a series of direct shear tests to study the effect of geogrid orientation and soil volume changes on shear strength of granular soil and suggested a numerical model to calculate the shear strength increase in the reinforced soil mass which was verified by experimental results. The model proposed was a variation of the model proposed by Palmeira (1987). For all the orientations of the reinforcement, the reinforced soil exhibited an intercept with the shear stress axis. This intercept has been termed by Yang (1972) as "apparent cohesion" and by Schlosser and Long (1974) as "additional confinement".

Large size direct shear test can, although, better simulate the field conditions compared to the conventional direct shear apparatus, it has verified that the scale was not really a problem (Palmeira 1987) and that the difference in the results using the two devices was not much noteworthy (Vieria et al. 2013). Palmeira (1987) constructed a large-scale direct shear apparatus and tested a cubic soil sample of unit dimensions 1000 mm. He demonstrated that the increase in the strength of reinforced soil depended upon the form, type and mechanical behavior of reinforcement. While studying the effect of scale on the frictional behavior of sand-geomembrane interfaces, it was concluded that large size shear box overestimates the shear strength of the interface compared to small scale direct shear device (Hsieh and Hsieh 2003; Vieria et al. 2013). Similar experimental studies on the mechanism and strength characteristics of reinforced soil using tension resisting elements have also been reported by various other researchers (e.g. Talwar 1981; Gray and Ohashi 1983; Shewbridge and Sitar 1989; Singh 1991; Michalowski and Cermak 2002; Ranjan et al. 1996; Rao et al. 2006; Babu and Vasudevan 2008; Saran 2010; Shukla et al. 2010; Singh et al. 2011; Vieria et al. 2013; Mir 2014; Shukla and Jha 2014; Infante et al. 2016).

In addition to experimental studies several numerical models of geosynthetic materials developed for the study of soil-reinforcement interaction can be found in the literature (Perkins 2000; Perkins and Edens 2003; Sharma et al. 2009; Ezzein et al. 2015; Belheine et al. 2009) most which are essentially based on simplifying assumptions like modeling the geogrid as a planer sheet. Perkins (2000) used elastic-plastic-creep model for modeling geosythetic and elastic-perfectly plastic geosynthetic-soil interaction model. The model predictions were found reasonably in agreement with the results of a series of pullout tests performed by the author. The planer sheet assumption does not take into account the geometric parameters (like aperture size, etc.) of reinforcement which are responsible for soil-geogrid interlocking. Several non-linear finite element and discrete element models proposed by various researchers (Cundall and Strack 1979; McDowell et al. 2006; Belheine et al. 2009; Hussein and Meguid 2016) considered the modeling of soil and geogrid in 3-dimensional geometry using Discrete Element Method (DEM). The discontinuous nature of soil was rationally taken into account using discrete element modelling. However, the continuity of the geogrid was not fully apprehended using DEM. Tran et al. (2013) carried out a 3D finite element analysis to model pullout loading test. Use of FEM for modelling geosynthetic materials and DEM for modelling the soil is a much reasonable approach for studying the soil-geogid interaction. Tran et al. (2015) numerically studied the 3D geogrid-soil interaction using FE-DE Framework. The soil was modelled using Discrete Elements (DEs) while as geogrid was modelled using Finite Elements (FEs). The FE-DE approach effectively captured the detailed 3D response of both geogrid and soil backfill in the soil-geogrid interaction. Apart from direct shear tests and pullout load test, soil-reinforcement interaction studies have been done by many researchers using triaxial tests (Parihar et al. 2015; Peng and Zornberg 2016). Zornberg (2007) experimentally validated the discrete approach of fibre-reinforced soil using triaxial test results. Zhang et al. (2007) introduced 3D reinforcing element in sand and compared the strength improvement using triaxial tests with horizontally reinforced sand. The experimental results were used to analyse a retaining structure reinforced with 3D reinforcement using finite element analysis. However, there is not much specific literature available on how stiffness of reinforcement affects the type of major reinforcement force developed at various orientations in reinforced soil and therefore, authors of this paper have made an attempt to evaluate the effect of stiffness and orientation of reinforcement on the shear strength of sand.

In this study, a series of small size direct shear tests were conducted on clean, poorly graded medium grained sand under undrained condition. The two type of reinforcements were used of relatively varying stiffnesses. The sand was tested in loose and medium dense states with reinforcement orientation (θ) varying from $-90°$ (inwards) to $+90°$ (outward) with vertical. The reinforcements were only embedded in the sand and were not anchored to the shear box at any point. Also, the post peak behavior of the reinforced soil at different orientations and normal stress values was studied with both types of reinforcement. The conclusions derived from the investigation suggested that the attainment of maximum increase in shear strength was observed for specific (optimum) orientations with the vertical for inward and outward orientations and that the stiffness of reinforcement influenced the optimum orientation of reinforcement corresponding to the attainment of maximum shear strength in terms of the composite friction angle as well as the apparent cohesion developed due to reinforcement in the reinforced soil mass.

1.1 Objective of Current Study

The current study is aimed to investigate the load transfer mechanism between the reinforcements of varying stiffnesses and the sand for different reinforcement orientations using small size direct shear tests. The objectives of the current study are briefly described as:

1. To evaluate the shear strength of unreinforced and reinforced sand using small size direct shear apparatus,
2. To study the effect of stiffness of reinforcement on the development of major reinforcement force, and
3. To study the effect of orientation of reinforcement on the development of major reinforcement force on the given sand.

2 Materials and Methodology

2.1 Properties of Sand

The standard test sand used throughout the test program was classified as poorly graded sand (SP) with properties shown in Table 1. The particle size distribution curve for the sand is shown in Fig. 2. All the tests have been conducted as per relevant Codal procedures (IS: 1498; IS: 2720).

2.2 Properties of Reinforcement

Two types of reinforcements of annealed rigid reinforcement binding wire and flexible geogrid (SGi-100) strips were used of relatively varying stiffnesses. For use as rigid reinforcement, 20 mm long black annealed reinforcement binding wire was used which

Table 1. Physical Properties of sand

Material properties		Test values
Specific Gravity		2.66
Void ratios	e_{max}	0.74
	e_{min}	0.55
Maximum unit weight, γ_{max} (kN/m^3)		16.9
Mean Particle size, D_{50} (mm)		0.78
Coefficient of Uniformity, C_u		3.15
Coefficient of Curvature, C_c		1.84
Particle size range, (mm)		<0.75–1.18

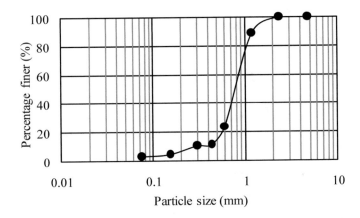

Fig. 2. Particle size distribution curve for sand sample

was glued with sand particles to provide the required friction giving an effective diameter of 1.63 mm (Fig. 3a). Another reinforcement that was used as flexible reinforcement was a 25 mm long, 10 mm wide and 2.5 mm thick flexible geogrid strips cut out from Geogrid-SGi100 (Fig. 3b). Geogrid-SGi 100 is polyester (PET) geogrid reinforcement for soil, which has a Tensile strength (ASTM D 6637 – Method A ASTM D4354; ASTM D 5321; IS: 13325) of 100 kN/m in Machine Direction-MD and 30 kN/m in Cross Machine Direction-CMD (Fig. 3c) with aperture size 29 mm × 62 mm. Rigid reinforcement was used as 3 in number, embedded centrally in a single row perpendicular to the direction of shear. While the flexible reinforcement used was 1 in number, centrally placed in the shear box.

2.3   Test Setup and Sample Preparation

To investigate the interaction between sand and the reinforcement, small size direct shear apparatus was used (Fig. 4). The reinforced samples were prepared in two ways: by driving and by placing. For reinforcement binding wire, the reinforcement was

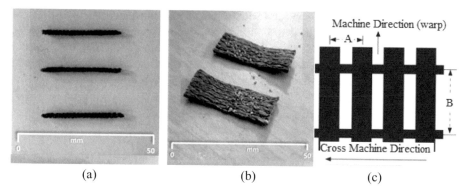

Fig. 3. Reinforcements used; (a) rigid reinforcement, (b) flexible reinforcement, (c) Geogrid rib formation during manufacturing process

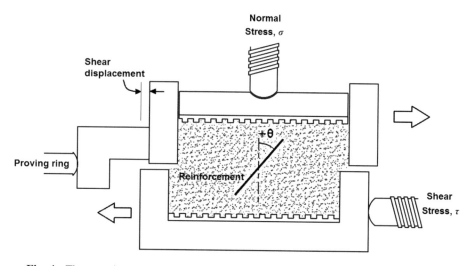

Fig. 4. The experimental setup showing the reinforcement at angle $+\theta$ with the vertical

driven into the sand placed (and compacted to the required density) in the lower half of the shear box at the required angle, and then the remaining sand was poured into the shear box and compacted as per the requirement. In case of geogrid strips, the method of placing the reinforcement at the time of sample preparation was adopted. The reinforcement was maintained at required angle by temporary support till the soil was filled around the reinforcement in the lower half of shear box and compacted to the required density. The support was then removed and the sand was poured in the upper half of the shear box and compacted as per the requirement. After preparation of geogrid reinforced sand samples, Direct shear tests were conducted using conventional direct shear box as per Standard codal procedures (IS 2720 – Part 13; ASTM D 5321).

2.3.1   Sign Convention Used for Orientation of Reinforcement

The reinforcement was oriented in both compressive and tensile strain zones. Orientation of reinforcement, theta (θ), is measured with vertical, i.e. normal to shear plane as shown in Fig. 5. Angle of orientation (θ) is taken as positive when reinforcement is in tension zone while it is taken as negative when placed in compression.

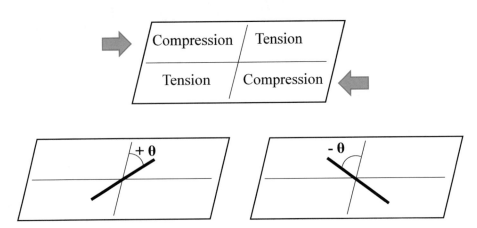

Fig. 5. Sign convention used for the reinforcement orientation

3   Results and Discussions

3.1   Unreinforced Test Results

The shear stress parameters of unreinforced soil were first determined for three different densities; loose ($R_D = 25\%$), medium dense ($R_D = 50\%$) and dense ($R_D = 75\%$). The values of test void ratios (e_{test}), relative densities (I_D), cohesion intercepts (c) and internal friction angles (ϕ) for the three densities are given in Table 2. Figure 6 shows the Mohr failure envelope for the loose, medium and dense states of unreinforced sand.

Table 2. Shear parameters of unreinforced sand

Density	e_{test}	I_D (%)	c (kPa)	$\phi°$
Loose	0.69 ± 0.02	25%	9	36.5
Medium	0.64 ± 0.02	50%	6.4	40.6
Dense	0.59 ± 0.02	75%	3.8	44.3

3.2   Reinforced Test Results Using Rigid Reinforcement

Direct shear tests were conducted on reinforced sand for reinforcement orientations, $\theta = 0°$, $15°$, $30°$, $45°$, $60°$ and $90°$ with normal to shear plane and vertical loads $\sigma_v = 50$, 100 and 150 kPa. The undrained shear parameters obtained are presented in tabular form in Table 3. The variation of apparent cohesion due to reinforcement (c_R)

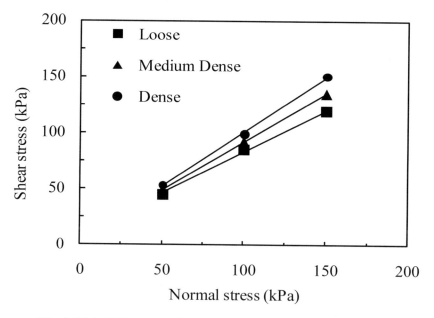

Fig. 6. Mohr Failure Envelope for various densities of unreinforced sand

Table 3. Undrained Shear parameters of reinforced medium dense sand for various orientations

θ	0°	15°	30°	45°	60°	90°
c_R (kPa)	10	17.5	15.0	12.5	13.8	5.1
$\phi°$		38.6	40.2	40.5	40.2	39.6	41.2

with the orientation of reinforcement (θ) initially shows an increasing and then a decreasing trend with a peak value at around 30°. Variation of composite friction angle (ϕ_c) with the orientation of reinforcement (θ) also shows a similar behavior, i.e. peak at around 30°, but shows an exception value for $\theta = 90°$. This may be due to the reason that the reinforcement in this orientation is parallel to and lies in the shear plane itself, this may result in greater interlocking between sand and the sand roughened rein-forcement explaining the exceptionally higher value of friction angle for this orienta-tion. The variation of shear strength as a whole with the orientation of reinforcement shows a peak between orientations of $\theta = 25°$ and $\theta = 30°$ as shown in Fig. 7. Boundary data from reinforced test on medium dense sand using rigid reinforcement at various orientations (only $\theta = 15°$, 30°, 45°, 90° shown) are presented in Fig. 8 in the form of load/stress-displacement graph with the corresponding unreinforced test at the same relative density and applied vertical stress of $\sigma_v = 50$ kPa. The vertical axis of the graph shows the normalized shear stress (τ_f / σ_v), also called the stress ratio as a function of horizontal/shear displacement.

For all the orientations of the reinforcement, the reinforced sand shows higher peak strength than the unreinforced sample, however, the behavior was slightly different for

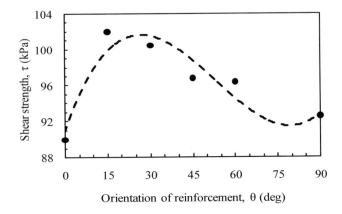

Fig. 7. Variation of shear strength of reinforced sand with the orientation of reinforcement (σ = 100 kPa)

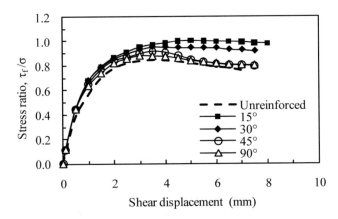

Fig. 8. Boundary data from reinforced direct shear test on medium dense sand

various orientations. Initially, with increasing shear displacement there is a little difference in the mobilized shear strength between the reinforced and unreinforced samples. For orientations, $\theta \geq 45°$ with the vertical, as shear displacement increases, the measured strength goes above the measured strength of unreinforced sample. After reaching a maximum value, strain softening occurs until the residual strength is reached, like for unreinforced sample. For orientations $\theta < 45°$ with the vertical, the strength increases to a maximum with no strain-softening taking place thereafter as was in case of unreinforced sample. Bauer and Zhao (1993) also observed similar behavior of reinforced samples at various orientations. The shear box used, however, was a large direct shear apparatus (1000 mm × 1000 mm × 940 mm) and the reinforcement used was a flexible geogrid which was not embedded in the sand sample but was anchored to the shear box at both top and bottom ends.

For the reinforced tests conducted on loose sand, the reinforcement was oriented at $\theta = -90°$, $-60°$, $-30°$, $0°$, $15°$, $30°$, $60°$ and $90°$ with normal to shear plane and under a vertical load of $\sigma_v = 100$ kPa. The variation of increase in shear strength of sand in terms of stress ratio (τ_{ext}/σ_y) with the reinforcement orientation (θ) for loose sand is shown in Fig. 9. The figure also shows variation as observed by Jewel and Wroth (1987) for a sheet of rough reinforcement placed at different orientations in the sand, for comparison. It is important to mention that their observation data has been scaled down by a factor of 0.03 for better comparison. Here, τ_{ext} represents the difference between the mobilized shear strength in reinforced sample and the unreinforced one for the same normal stress σ_y. The graph shows similar variation as observed by Jewel (1980). It shows a peak value near $+30°$ orientation.

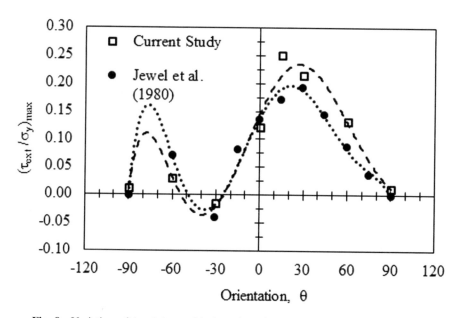

Fig. 9. Variation of $(\tau_{ext}/\sigma)_{max}$ with the orientation of reinforcement in loose state

3.3 Reinforced Test Results Using Flexible Reinforcement

Increase in the strength of the reinforced soil when represented in the form of shear stress vs. normal stress graph, (i.e. the Mohr-failure envelops) of the reinforced soils at different orientations shows a notable trend as observed in Fig. 10. It can be observed from Fig. 8 that the failure envelops of reinforced samples with positive orientation of reinforcement are similar (in terms of slope) to the failure envelop of unreinforced sample, and are shifted upwards signifying a higher shear strength of soil for all the values of normal stress than the unreinforced sample. However, the reinforced samples having negative orientation of reinforcement have a flatter failure envelop than that of the unreinforced sample. This signifies that the shear strength of sand having such orientation of reinforcement will show higher strength at lower normal stresses while

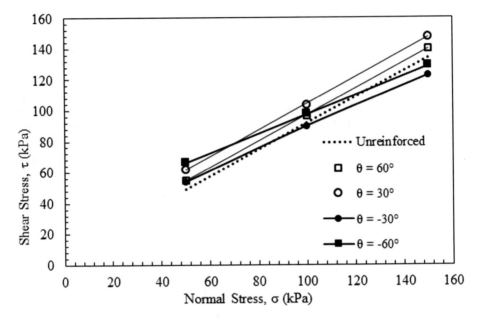

Fig. 10. Mohr-failure envelops of the reinforced medium dense soil at different orientations using flexible reinforcement

the strength becomes less than their unreinforced counterparts at higher values of normal stress. It may be noted that $\theta = + 30°$ nearly coincides with the minor principle stress plane while $\theta = -60°$ nearly coincides with the major principle stress plane of the unreinforced sand.

4 Conclusions

Soil-reinforcement interaction studies reported in literature are mostly concerned with the use of reinforcement in axial tension. For the materials which do not possess much flexural stiffness, any increase in shear resistance is attributed to the additional tensile strain mobilized in the reinforcement. However, in case of rigid reinforcement, in addition to the axial stresses developed in these elements there is a greater chance of attributing the strength developed in the soil mass to the bearing resistance of the reinforcement due to its flexural stiffness. Direct shear tests were performed to analyze the behavior of poorly graded medium grained sand reinforced using reinforcements of relatively varying stiffnesses. Based on the interpretation of the test results, the following conclusions can be drawn:

1. The orientation of the reinforcement with respect to shear plane has a significant effect on the shear resistance of the soil.
2. The attainment of maximum increase in shear strength, in general, was observed when the reinforcement was oriented at about +30° with the vertical, which more or

less coincides with the direction of principle tensile plane, and about $-60°$ with the vertical for out ward orientation.

3. The stiffness of reinforcement is a decisive factor in the determination of optimum orientation of reinforcement which corresponds to attainment of overall maximum shear strength.

4. At lower normal stress levels in case of flexible reinforcement, orientation of reinforcement along the direction of major principle stress plane gives higher strength than the orientation along the minor principle stress plane. However, for higher stress levels, use of reinforcement in the direction of major principle stress plane reduces the shear strength beyond that of the unreinforced sand.

It can also be concluded that the use of small size apparatus was feasible as the results obtained were in agreement with the previous findings in the literature so the conclusions derived may be considered reliable.

Acknowledgments. The investigation reported in this paper forms a part of the Research Project work on reinforced soils at NIT Srinagar. Thanks are due to Faculty of Geotechnical Engineering Division and supporting staff of the Soil Mechanics laboratory for their timely help during the course of investigation.

References

ASTM D 5321. Standard test method for determining the coefficient of soil and geosynthetic or geosynthetic and geosynthetic friction by direct shear method. ASTM Designation: ASTMD5321.08, ASTM (2008)

ASTM D 6637. Standard test method for determining tensile properties of geogrids by the single or multi-rib tensile method. ASTM International, West Conshohocken, PA. www.astm.org (2015)

ASTM D 4354. Standard practice for sampling of geosynthetics and rolled erosion control products (RECPs) for testing. ASTM International, West Conshohocken, PA. www.astm.org (2012)

Babu Sivakumar, G.L., Vasudevan, A.K. Strength and stiffness response of coir -reinforced tropical soil. J. Mater. Civ. Eng. ASCE **20**(9), 571–578 (2008)

Bauer, G.E., Zhao, Y.: Evaluation of shear strength and dilatancy behavior of reinforced soil from direct shear tests. In: Jonathan Cheng, S.C. (ed.) Geosynthetic soil Reinforcement Testing Procedures, ASTM STP 1190, American Society for Testing and Materials, Philadelphia (1993)

Belheine, N., Plassiard, J.P., Donzé, F.V., Darve, F., Seridi, A.: Numerical simulation of drained triaxial test using 3D discrete element modeling. Comput. Geotech. **36**(1–2), 320–331 (2009)

Cundall, P.A., Strack, D.L.: A discrete numerical model for granular assemblies. Geotechnique **29**(1), 47–65 (1979)

Dyer, M.R.: Observation of the stress distribution in crushed glass with applications to soil reinforcement. Ph.D. thesis, Magdalen College, Michaelmas Term

Ezzein, F.M., Bathurst, R.J., Kongkitkul, W.: Nonlinear load–strain modeling of polypropylene geogrids during constant rate-of-strain loading. Polym. Eng. Sci. **55**, 1617–1627 (2015)

Gray, D.H., Ohashi, H.: Mechanics of fibre reinforcing in sand. J. Geotech. Eng. ASCE **112**(8), 335–353 (1983)

Horpibulsuk, S., Niramitkornburee, A.: Pullout resistance of bearing reinforcement embedded in sand. Soils Found. **50**(2), 215–226 (2010)

Hsieh, C., Hsieh, M.W.: Load plate rigidity and scale effects on the frictional behavior of sand/geomembrane interfaces. Geotext. Geomembr. **21**, 25–47 (2003)

Hsieh, C.W., Chen, G.H., Wu, J.H.: The shear behavior obtained from the direct shear and pullout tests for different poor graded soil-geosynthetic systems. J. GeoEng. **6**(1), 15–26 (2011)

Hussein, M., Meguid, M.A.: A three-dimensional finite element approach for modeling biaxial geogrid with application to geogrid-reinforced soils. Geotext. Geomembr. **44**(3), 295–307 (2016)

Infante, D.J.U., Martinez, G.M.A., Arrua, P.A., Eberhardt, M.: Shear strength behavior of different geosynthetic reinforced soil structure from direct shear test. Int. J. Geosynth. Ground Eng **2**, 17 (2016). https://doi.org/10.1007/s40891-016-0058-2

IS: 1498: Method of Test on Soils: Classification and Identification of Soils for General Engineering Purposes. Bureau of Indian Standards, New Delhi (1970)

IS: 2720-Part. 4: Method of Test on Soils: Grain Size Analysis. Bureau of Indian Standards, New Delhi (1985)

IS: 2720-Part. III: Method of Test on Soils: Determination of Specific Gravity of Fine, Medium and Coarse Grained Soil. Bureau of Indian Standards, New Delhi (1980)

IS: 2720-Part.14. Method of Test on Soils: Determination of Density Index (R.D) of Cohesion Less Soil. Bureau of Indian Standards, New Delhi (1983)

IS: 2720 – Part 13: Method of Test on Soils: Direct Shear Test for Soils. Bureau of Indian Standards, New Delhi (1986)

IS: 13325: Methods of Test for Soils: Determination of Tensile Properties of Extruded Polymer Geo-Grids Using the Wide Strip - Test Method. Bureau of Indian Standards, New Delhi (1992)

Jewell, R.A.: Some effects of reinforcement on the mechanical behavior of soils. Ph.D. thesis, University of Cambridge (1980)

Jewell, R.A., Wroth, C.P.: Direct shear tests on reinforced sand. Géotechnique **37**(1), 53–68 (1987). https://doi.org/10.1680/geot.1987.37.1.53

Juran, I., Guermazi, A., Chen, C.L., Ider, M.H.: Modelling and simulation of load transfer in reinforced soil: Part 1. Int. J. Numer. Anal. Methods Geomech. **12**, 141–155 (1988)

Liu, C.-N., Ho, Y.-H., Huang, J.-W.: Large scale direct shear tests of soil/PET-yarn geogrid interfaces. Geotext. Geomembr. **27**(1), 19–30 (2009)

McDowell, G.R., Harireche, O., Konietzky, H., Brown, S.F., Thom, N.H.: Discrete element modelling of geogrid-reinforced aggregates. Proc. Inst. Civ. Eng. Geotech. Eng. **159**(1), 35–48 (2006)

McGown, A., Andrawes, K.Z., Al-Hasani, M.M.: Effect of inclusion properties on the behavior of sand. Geotechnique **28**(3), 327–346 (1978)

Michalowski, R.L., Cerma'k, J.: Strength anisotropy of fibre reinforced sand. Comput. Geotech. **29**(4), 279–299 (2002)

Mir, B.A.: Geosynthetics applications in highway construction in J&K: Sustainable infrastructure development. In: Proceedings of National Conference "GEPSID", Ludhiana, India, pp. 10–20 (2014)

Mofiz, S.A., Taha, M.R., Sharker, D.C.: Mechanical stress-strain characteristics and model behavior of geosynthetic reinforced soil composites. In: 17th ASCE Engineering Mechanics Conference, EM 2004, pp. 1–8. University of Delaware, Newark (2004)

Palmeira, E.M.: The Study of Soil - reinforcement interaction by means of large scale laboratory tests. Ph.D. thesis, University of Oxford, England (1987)

Parihar, N.S., Shukla, R.P., Gupta, A.K.: Effect of reinforcement on soil. Int. J. Appl. Eng. Res. 10(55), 4147–4151 (2015). ISSN 0973-4562

Peng, X., Zornberg, J.G.: Evaluation of load transfer in triaxial geogrids using transparent soil. In: Proceedings of the 3rd Pan-American Conference on Geosynthetics, Miami, Florida, 10–13 April, vol. 2, pp. 1520–1531 (2016)

Perkins, S. W.: Constitutive modeling of geosynthetics. Geotext. Geomembr. 18(5) 273–292 (2000), ISSN 0266-1144. https://doi.org/10.1016/S0266-1144(99)00021-7

Perkins, S.W., Edens, M.Q.: Finite element modeling of a geosynthetic pullout test. Geotech. Geol. Eng. 21, 357e375 (2003)

Ranjan, G., Vasan, R.M., Charan, H.D.: Probabilistic analysis of randomly distributed fibre-reinforced soil. J. Geotech. Eng. ASCE 122(6), 419–426 (1996)

Rao, A.S., Rao, K.V.N., Sabitha, G., Surest, L.K.: Load deformation behavior of fibre-reinforced gravel beds overlying soft clay. In: Proceedings of National Conference on Corrective Engineering Practices in Troublesome Soils (CONCEPTS), Kakinad4, 8–9 July 2006, pp. 187–190 (2006)

Saran, S.: Reinforced Soil and Its Engineering Applications. I.K. International Publishing House Pvt. Ltd., New Delhi (2010)

Shewbridge, S.E., Sitar, N.: Deformation characteristic of reinforced sand in direct shear. J. Geotech. Eng. ASCE 115(8), 1134–1147 (1989)

Sharma, R., Chen, Q., Abu-Farsakh, M., Yoon, S.: Analytical modeling of geogrid reinforced soil foundation. Geotext. Geomembr. 27(1), 63–72 (2009). ISSN 0266-1144

Shukla, S.K., Shivakugan, N., Singh, A.K.: Analytical model for fiber-reinforced granular soil under high confining stress. J. Mater. Civ. Eng. ASCE 22(9), 935–942 (2010)

Shukla, S.K., Jha, J.N.: Fibre-reinforced soils – Basic developments. In: Proceedings of National Conference "GEPSID", Ludhiana, India, pp. 1–9 (2014)

Singh, V.K.: Strength and deformation characteristics of sand reinforced with geotextiles. M.E. dissertation, IIT, Roorkee, India (1991)

Singh, H.P., Sharma, A., Chanda, N.: Study of strength characteristics of coir fiber-reinforced soil. In: International Conference on Advances in Material and Techniques for Infrastructure Development (AMTID 2011), held at NIT Calicut Kerala, India. Paper No. G002, 28–30 September 2011

Schlosser, F., Long, N.T.: Recent results in French research on reinforced earth. J. Constr. Div. 100, 223–237 (1974). American Society of Civil Engineers, Reston

Vieria, C.S., Lopes, M.L., Caldeira, L.: Soil-geosynthetic interface shear strength by simple and direct shear tests. In: Proceedings of the 18th International Conference on Soil Mechanics and Geotechnical Engineering, Paris (2013)

Talwar, D.V.: Behavior of reinforced earth in retaining structures and shallow foundations. Ph.D. thesis, IIT, Roorkee, India (1981)

Tran, V., Meguid, M.A., Chouinard L.E.: A finite-discrete element framework for the 3D modeling of geogrid-soil interaction under pullout loading conditions. Geotext. Geomembr. 37, 1–9 (2013)

Tran, V., Meguid, M.A., Chouinard, L.E.: Three-dimensional analysis of geogrid reinforced soil using finite-discrete element framework. ASCE's Int. J. Geomech. 15(4), 1–19 (2015)

Yang, Z.: Strength and deformation characteristics of reinforced sand. Ph.D. thesis, University of California at Los Angeles, Los Angeles, CA (1972)

A Review on Improvement of Subgrade Soil Using Coir Geotextiles

Sridhar Rajagopalaiah[✉]

Sri Venkateshwara College of Engineering, Bangalore, Karnataka, India
sridharrajagopalg@gmail.com

Abstract. India has one of the largest road networks in the world, aggregating to about 33 lakh km at present. However many of the existing roads are becoming structurally inadequate because of the rapid growth in traffic volume and axle loading. At locations with adequate subgrade bearing capacity/CBR value, a layer of suitable granular material can improve the bearing capacity to carry the expected traffic load. But at sites with CBR less than 2%, problems of shear failure and excessive rutting are frequently encountered. The ground improvement alternatives such as excavation and replacement of unsuitable material, deep compaction, chemical stabilization, pre loading and polymeric geosynthetics etc., are often used at such sites. The cost of these processes as well as virgin material involved is usually high and as such they are yet to be commonly used in developing nations like India. In this situation natural fiber products hold promise for rural road construction over soft clay. The Review has been made on the performance coir geotextile as a reinforcement in various soil conditions.

Keywords: Coir geotextile · Bearing capacity · Subgrade · Rutting

1 Introduction

Some policies were implemented for period of 60 years from the governments for the development of rural roads. Government of India had launched Pradhan Mantri Gram Sadak Yojana (PMGSY) in the year 2000 to provide connectivity to unconnected eligible habitations and upgradation of select existing roads to the standards. PMGSY is being implemented since the Year 2000. The projects for 1,09,010 habitations have been sanctioned out of total 1,36,464 eligible habitations by clearing the proposals for 4,20,637 kms roads. In the Year 2003–04, detailed District Rural Road Plans were made in every district of the country and Core Networks to ensure single connectivity to all habitations eligible under the programme were identified. But even today there is a requirement of further plans and implementations. The Rural Roads Vision – 2025 targeting at implementation of all weather road access to all habitations by the year 2025. Engineers are often experiencing with the difficulty of constructing roadbeds on or with soils which do not have adequate strength to support wheel loads upon them either in construction stage or during the life of the pavement. Road construction and maintenance along soils have been problematic due to their inherent potential for volume change in the presence of water, which adversely impacts the performance of roads.

© Springer Nature Switzerland AG 2019
M. Meguid et al. (Eds.): GeoMEast 2018, SUCI, pp. 152–164, 2019.
https://doi.org/10.1007/978-3-030-01944-0_12

Unless the subgrade is appropriately treated at the construction stage, the total transportation cost will increase substantially due to deteriorated pavement performance and associated road user costs. Geosynthetics have proven to be the most versatile and cost effective ground modification material. Most of the geosynthetics are made of polymeric material. Natural geotextiles made of coir, jute, etc. are preferable to synthetic fibres on account of the fact that the material is environment friendly and ecologically compatible as it gets degraded with time. Moreover, natural fibres are less costly which make it a better choice compared to synthetic fibres. Geotextiles are permeable fabrics which, when used in connection with soil, have the ability to separate, filter, reinforce and drain. It is not only allows lessening in the thickness of the pavement on a soft subgrade by the reinforcement action of geotextiles but gives less maintenance problems for long-term use. Coir geotextiles are made from coconut fibre which is a natural material composed of ligon cellulose cell derived from the husk of coconut. Reinforcing the sub grade using coir fabrics helps to distribute the load over a wider area and results in enhanced bearing capacity and reduced settlement. In the case of a pavement, this would result in reduced thickness of road structure and less earth work when it is used as a membrane between the subgrade and the overlying thin granular sub base layer. For unstable and wet subgrade, a coir fabric appears to provide satisfactory solution to stability and drainage problems. Coir geotextile for Strengthening Subgrade Coir is a 100% organic naturally occurring fibre, from a renewable source obtained from coconut husk. Naturally resistant to rot, moulds and moisture, it not only reduces the thickness of the pavement but also chemical treatment is not needed. Geotextiles made of coir are ideally suited for low cost applications because coir is available in India in abundance at very low price compared to other synthetic geotextiles. These geotextiles can be applied in the construction of unpaved roads where they can be effectively serving the purposes of reinforcement, separation, filtration and drainage. Coir geotextiles are found to last for four to six years within the soil environment depending on the physical and chemical properties of the soil. When it is used as reinforcement, the coir layers can share the load with soil until its degradation thus increasing the load bearing capacity of the subgrade. When coir geotextiles are used, they also serve as good separators and drainage filters. In many instances, the strength of subgrade soil increases in course of time as the soil undergoes consolidation induced by the traffic loads. At this stage, the subgrade may be strong enough to support the loads on its own without the necessity for reinforcement. For such applications, where the strength of subgrade increases with elapsed time, the natural reinforcement products are extremely suitable. After the degradation of the coir geotextiles, the organic skeleton remains in place in compressed form which will act as a filter cake keeping the moisture content of the subgrade soil constant.

But these additives do not mix properly with soil. Geotextiles form one of the largest groups of geosynthetics. One of the most popular applications of Geotextiles is in the construction of pavements and embankments on soft soil. They are indeed textiles in the traditional sense, consisting mainly of synthetic fibers, though natural fibers are also used for manufacturing. They can be woven or non-woven type (Subaida et al. 2009). There are enormous specific application areas for geotextiles. Generally the fabric performs at least one of the four discrete functions viz., separation, reinforcement, filtration, drainage (IRC: SP: 59-2002). The low cost of natural fibers, the growing concern over the impact of the use and disposal of synthetic materials has recently led to a renowned interest in the possible advantages of natural geotextiles. Natural geotextiles made of coconut fiber, jute fiber; sisal, etc. can be used as an alternative to polymeric geosynthetic materials (Ling and Liu 2001). Coir net is readymade material, cheap, easy laying in field and biodegradable. Coir geotextiles find application in a number of situations in geotechnical/ highway engineering practice. Coir geotextile is used for strengthening subgrade, it can be used as an overlay or interlay, the former protecting the surface from runoff and the latter performing the functions of separation, reinforcement, filtration and drainage.

2 Materials

Coir Geotextile. Coir geotextiles with its Indian notation "Coir Bhoovastra", a generic member of the geosynthetic family, are made from the coconut fibre extracted from the husk of the coconut fruit. Coir geotextile is shown in Fig. 1(a) and typical application of coir geotextile along embankment is shown in Fig. 1(b).

(a) Coir Geotextile   (b) Typical Application of Coir Geotextile

Fig. 1. Coir Geotextile

Properties of Coir Geotextile. Physical Properties of coir geotextiles are Mass per unit area, Thickness, and Specific gravity. Mechanical Properties of coir geotextiles are strip tensile strength, wide width tensile strength, trapezoidal tear strength, grab tensile strength, drop cone penetration resistance, puncture resistance, burst strength, interface friction, Pull out resistance and Sewn seam strength.

3 Coir Geotextile as a Reinforcement Material

Mehndiratta et al. (1993) carried out a study based on the natural geo-textiles in highway embankments. In this study, they made an attempt to increase the life of coir as well as to reduce the microbial attack and faster degradation by using chemical agents. Figure 2 shows the placement of coir geotextile in typical pavement section.

Fig. 2. Location of coir geotextile in typical pavement section.

Babu et al. (2008) had formulated a new design methodology using IRC Guidelines, to reinforce the sub-grade using a natural geo-textile so as to improve the strength of sub-grade. In this paper, a design methodology using IRC Guidelines for the design of Coir geo-textile reinforced roads had been found out on the basis of laboratory experimental data and mathematical formulations when the CBR value is less than 2%. The method adopted was to reinforce the sub-grade using geo-textiles. Babu et al. (2008) had conducted studies based on the load deformation behaviour of unpaved roads due to the placement of coir geotextiles on the basis of certain laboratory investigations. The four fundamental reinforcement mechanisms such as separation, lateral restraint, improved bearing capacity and tensioned membrane effect has been identified which reveals the use of geotextiles in unpaved road construction. The experiments were conducted by using commercially available three varieties of coir geotextiles (two woven type designated as H2M6, H2M8 and one non-woven type, AGLC/201) placed at sub-grade sub-base interface and also between layers of subbase. Babu et al. (2008) had reported the results of exhaustive study carried out to explore the behaviour of coir geotextile reinforced subgrade soils in terms of California Bearing Ratio. From the studies, it is clear that the existence of coir geotextiles influences the strength of subgrade due to the interaction between soil and coir geotextile in soaked and unsoaked condition. Rao and Balan (2000) had successfully used coir geotextles in limited applications of geotechnical engineering. This paper deals with an overview of various case studies and model studies conducted by the author on the various uses of coir geotextiles. It gives an overview on the potential uses of geotextiles in various civil engineering applications. Subaida et al. (2008) conducted experiments to investigate the beneficial use of coir geotextiles as reinforcing material in a two layer pavement section. The effects of placement position and thickness of geotextiles on the performance of reinforced sections were investigated using two base course thickness and two types of woven coir geotextiles. The test results indicate an enhancement in the bearing capacity of thin sections. Baruah et al. (2010) studied the variation in the resistance of soil in terms of CBR value with respect to placement of coir mat in CBR mould from top surface of soil. The test results revealed that, in soaked condition the inclusion of coir mat has improved the CBR by nearly two times. Raji et al. (2011) carried out field simulation on subgrade soil with different combination of fly ash, cement and coir geotextile using a wheel tracking apparatus. It was reported that the subgrade obtained with the application of fly ash, cement and coir geotextile gives significant strength and stability to the soil.

Subaida et al. (2008) conducted experiments to investigate the beneficial use of coir geotextiles as reinforcing material in a two layer pavement section. The effects of placement position and thickness of geotextiles on the performance of reinforced sections were investigated using two base course thickness and two types of woven coir geotextiles. Nithin et al. (2012) conducted model studies to investigate the beneficial use of coir geotextiles as reinforcing material on weak lateritic soil with Wet Mix Macadam (WMM) representing unpaved roads on soft subgrade. The coir geotextiles are kept at different levels in the subgrade model sections for studying the effect of position of geotextiles in upgrading the bearing capacity of soil under the monotonic loading system. It was reported that there is a considerable amount of increment of the bearing capacity of reinforced subgrade with respect to the unreinforced subgrade at a specified settlement. The placement of coir geotextiles at base and subgrade interface shows a significant increase in the load at higher settlements due to the membrane action. Whereas, placing the coir geotextile within the base course resulted in a considerable increase in load at small as well as at large settlements. Inclusion of a coir geotextile layer, at the interface of subgrade and subbase improves the load settlement characteristics and an additional layer of coir geotextile at mid-depth of subbase produces marginal improvement in the system. Provision of a layer of geotextile at the interface between subgrade and sub base reduces the deformation by 40%, which in turn results in the reduction of sub-base thickness required (Bhole et al. 2015). The improvement due to insertion of coir geotextile is shown in Fig. 3.

A geotextile is similar to a fabric. It is manufactured by interweaving together numerous yarns in a close-knit pattern. The pattern is tight enough to filter sand/aggregate particles, thus an apparent opening size (AOS) typically characterizes the openings of a geotextile A geosynthetic is affected by its surroundings or Environment. Environmental factors that contribute to the degradation of geosynthetics include UV radiation (sunlight), mechanical/physical wear, long duration loads, and temperature. For instance a polypropylene textile or grid, will creep when exposed to tensile loads. Creep is also enhanced by an increase in temperature and additionally, UV radiation in sunlight can cause serious degradation and weakening of polymer bonds.

There are many applications of geosynthetics. Even within the highway application of geosynthetics, further division is necessary for clarity. Geosynthetic highway applications can be split into two areas, which are unpaved and paved roads. It is

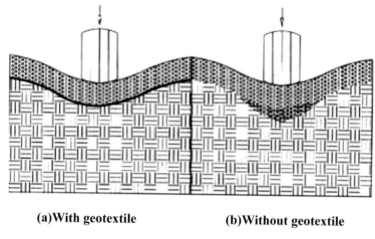

(a)With geotextile   **(b)Without geotextile**

Fig. 3. Effect of insertion of geotextile between the layers.

important Performance study on Coir Geotextiles in pavements having soft soil sub-grade to distinguish between the two, since different theories, physical mechanisms, design methodologies and failure criteria are utilized for each.

3.1 Unpaved Road- An Unpaved Road Haul Loads Across Undeveloped Terrain

Typically, such grades are crossed with a minimum amount of preparation that allows for an efficient movement of relatively few, but heavy, load repetitions. Rutting in the wheel paths is allowed but typically desired to be four inches or less in depth. Regrading or leveling of the ruts can be performed but is not typically, considered for an initial design of a layer of select granular material, which is placed upon the subgrade as a surface course. The purpose of this surface course is to transfer the surface load to the subgrade while spreading out the load to the subgrade, which effectively reduces the intensity of pressure on the subgrade (Bell and Steward 1977). A geosynthetic placed properly does improve an unpaved road. The most effective location of the geosynthetic is below the select granular material and on the subgrade surface (Das et al. 1998). In this location the geosynthetic provides separation, lateral restraint of the upper granular course and a tensioned membrane effect when strained extensively. A geotextile separates a granular course from a fine-grained subgrade, due to its relatively small apertures or apparent opening size (AOS). However, a geogrid also provides separation due to its less than 100% open area and better lateral restraint of upper granular particles. Due to interface friction and interlock with many individual ribs, a geogrid provides superior lateral restraint of the upper granular course, whereas the geotextile relies exclusively on interface friction for lateral restraint (Bell and Steward 1977). The tensioned membrane effect requires that the geosynthetic be extensively strained (i.e., deeply rutted) for this mechanism to contribute a significant benefit.

Rigarous analytical and experimental investigations concerning to geosynthetic interfaced aggregate – soil as unpaved road (Giroud and Noiray 1981; Giroud et al. 1985; Love et al. 1987; Burd 1995; Fannin and Sigurdsson 1996; Oloo et al. 1997; Som and Sahu 1999; Raymond and Ismail 2003; Watts and Blackman 2004; Giroud and Han 2004; Chew et al. 2005; Retzlaff et al. 2006; Kazimierowiicz-Frankowska 2007; Hu and Zhang 2007) may be accessed. Large-scale field tests and large scale laboratory tests (Elvidge and Raymond 1999; Bergado et al. 2001; Bhosale and Kambale 2008; Subaida et al. 2009) are investigated the working of unpaved roads. Laboratory CBR tests are done to study the use of natural coir and jute geotextile (Michael and Vinod 2009). CBR tests are also conducted by introducing geotextiles and geogrid in granular soil (Naeini and Mirzakhanlari 2008; Duncan-Williams and Attoh-Okine 2008; Dhule et al. 2011). Added to the above, based on CBR test, the influence of geotextile, geogrid and geonet are studied in clay with low or medium compressibility (Srivastava et al. 1995; Naeini and Moayed 2009; Nair and Latha 2010; Nair and Latha 2011) as soft subgrade in an unpaved road system.

3.2   Paved Road Performance Study on Coir Geotextiles in Pavements Having Soft Soil Subgrade

The other application is the paved road. This application also encompasses the unpaved application since during construction of a paved road relatively few repetitions of trucks heavily loaded with construction materials traverse the partially completed (unpaved) highway grade. This often leads the road to critical stage. Then, construction is completed with placement of an asphalt surface course, thus the highway is paved and open to the public. The opened highway is exposed to many repetitions from loaded truck traffic; however the intensity of subgrade load is considerably less due to the greater stiffness of the surface course. Benefits of an underlying geosynthetic during construction are apparent, but as time and greater numbers of load cycles pass, the benefits are not as clear for the paved road (Barksdale et al. 1989). Geogrids and geotextiles are the two types of geosynthetics most widely used in pavement systems at aggregate subgrade interface to reinforce or stabilize pavements. Field evidences suggest that both geogrid and geotextile could improve the performance of pavement sections constructed on weak soil. Several investigators have reported significant effects of pavement stabilization using geotextile reinforcement to improve the bearing capacity of subgrade soil. Maxwell et al. (2005) conducted a field demonstration to study how the performance of highway pavements is improved with geotextiles. In his research a field demonstration was conducted using a 21-m section along a Wisconsin highway (USH 45) near Antigo, Wisconsin, that incorporated three test sub-sections. Three different geosynthetics including a woven geotextile and two different types of geogrids had been used for stabilization. Observations made during and after construction indicate that all sections provided adequate support for the construction equipment and that no distress seems to be evident in any part of the highway. Large-scale experiments conducted on working platforms of crushed rock (breaker run stone or Grade 2 gravel) overlying a simulated soft subgrade. The tests were intended to simulate conditions during highway construction on soft subgrades Performance study on Coir Geotextiles in pavements having soft soil subgrade where the working platform is used to limit total deflections due to repetitive loads applied by construction traffic. Tests were conducted with and without geosynthetic reinforcement to evaluate how the required thickness of the working platform is affected by the presence of reinforcement. Working platforms reinforced by geosynthetics accumulated deformation at a slower rate than unreinforced working platforms, and in most cases deformation of the geosynthetic reinforced working platforms nearly ceased after 200 loading cycles. As a result, total deflections were always smaller (about a factor of two) for reinforced working platforms relative to unreinforced working platforms. Erickson and Drescher (2002) investigated the reinforcement function of geosynthetics for a typical Minnesota low volume roadways. From the study it was observed that the addition of a geosynthetic does provide reinforcement to the roadway as long as the geosynthetic is stiffer than the subgrade material. The service life of a roadway may also be increased with the addition of geosynthetic reinforcement. It was also observed that the deflection response of roadway is governed by the Young's modulus of the geosynthetic used. Since the deflections were controlled by the Young's modulus of the geosynthetic; the largest modulus geosynthetic produced the largest increase in service life. Schriver et al. (2002) conducted experimental study on geogrid reinforced lightweight aggregate beds to determine their subgrade modulus and increase in the bearing capacity ratio. From study it was observed that the geogrid reinforcement placed at sub base/aggregate interface effectively increases the service life of paved roads. Geogrid reinforcement provides a more uniform load distribution and a deduction in maximum settlement more at the asphalt-aggregate and aggregate-subgrade interface. Ranadive and Jadhav (2004) investigated the performance of geotextiles reinforcement in soil other than sand. In this study, model strip footing load tests are conducted on soil with and without single and multi-layers of geotextile at different depths below the footing. Testing was carried out on universal testing machine. From the study it was observed that bearing Performance study on coir geotextiles in pavements having soft soil subgrade capacity improved considerably for reinforced soil over unreinforced soil. To quantify the strength of reinforced soil the term bearing capacity ratio is usually used. It is the ratio of bearing capacity of reinforced soil to the bearing capacity of conventional soil. It was observed that for a single layer system, BCR (Bearing Capacity Ratio) for depth of layer below footing equal to 0.25B is maximum where B is the width of the footing and BCR decreases as the depth of layer increases and for multilayer system, BCR for a constant d/B ratio and S/B ratio, (where d is the depth of single reinforcing layer below footing and S is spacing between subsequent geotextile reinforcing layers when depth of top layer below footing was kept constant equal to 0.25B). The BCR is maximum for N = 4 but the percentage increase in BCR for N = 4 over BCR for N = 3 is very low. Thus N = 3 is recommended as optimum value. Investigators conducted plate load test to study the variation of load carrying capacity for both reinforced and unreinforced pavements. It was observed that the bearing capacity improved by providing coir geotextiles as reinforcement, reported an increase in bearing capacity by 1.83 times for reinforced pavement compared to unreinforced pavement. Venkatappa et al. (2005) conducted monotonic and cyclic load test on Kaolinite with geotextile placed at the interface of the two soils. It was found bearing pressure of the soil improved by about 33% when reinforced with coir geotextiles. Indian Roads Congress also suggest in its

Rural Road Manual (IRC: SP: 59-2002) the use of coir geotextile but no design methodology, construction guidelines and product specifications are mentioned.

From the CBR test results conducted by Baruah et al. (2010), it is observed that CBR value of the soil has improved upto 63% for unsoaked condition and 190% for soaked condition when the position of coir mat is at 1 cm from the top surface with the mould–plunger ratio of D/d = 3. The fact that in the soaked condition the inclusion of coir mat has improved the CBR by nearly two times has got special bearing to the state of Assam which is in general is a high rainfall and high water table state and roads have to perform very often in a soaked condition. It seems that coir mat reinforcement is almost imperative here. The CBR values decreases with the increase in mould size i.e. decrease in lateral confinement. The effect of lateral confinement is more for D/d = 2 and CBR values observed are also higher than others. For D/d = 5 results are almost identical with the field CBR test results. Thus it can be concluded that by increasing D/d ratio i.e. decreasing the lateral confinement, Investigators approached field conditions. At D/d = 5 we get field condition. Again the higher CBR values are obtained for position of coir mat at 1 cm. From the design aspects it is observed that the thickness of pavement may be reduced by approximately 75% if coir mat is placed above the subgrade.

Azeez et al. (2015) found that the use of coir geotextiles as subgrade reinforcement increased the subgrade strength and reduced the settlement. The coir geotextile significantly decreased the permanent vertical deformation over the loaded area of the pavement under repeated loading by restraining the lateral spreading of base material. Kumar and Rajkumar (2012) studied the performance of woven and nonwoven geotextile, interfaced between soft subgrade and Sasi and Aparna Sai (2017) unbound gravel in an unpaved flexible pavement system is carried out experimentally, utilizing the California Bearing Ratio (CBR) testing arrangement. And found that woven geotextiles give better performance compared to non-woven geotextiles. The CBR value of reinforced soil is increased compared to unreinforced soil. Thus by providing coir geotextiles the weak soil can be used for pavement construction. The maximum improvement in CBR value was obtained when the geotextile is positioned at 1/3H. Sodium hydroxide treated coir geotextiles gives better performance compared to untreated geotextiles. Hence treatment of coir geotextile is beneficial for improving the CBR value and the percentage improvement in CBR value is 66.8% when it is placed at 1/3H. The soaked CBR value also improved when coir geotextiles are used.

3.3   Use of Coir Geotextile in Expansive Soil Subgrade

The roads laid on BC soil bases develop undulations at the road surface due to loss of strength of the sub-grade through softening during monsoon. BC soil is a highly clayey soil and problem for highway engineers. In dry state it shrinks and becomes so hard that the clods cannot be easily pulverized for treatment for its use in road construction. However, when it is wet during rains, it swells as well as loses strength and poses serious problem with regard to subsequent performance of the road. All this results cracking in roads and for this is the reason that the road engineers do not prefer to construct roads on BC soil but have no option as black cotton soil is available in about one third part of the country, particularly in Madhya Pradesh. It is often impossible to build a stable base course over soft subgrade, without losing expensive base material which penetrates into the soft sub grade soil and hence a ground improvement method has to be resorted to. There are many methods of ground improvement such as cement stabilization, lime stabilization, chemical stabilization etc.

The Black cotton soils (BC) of India are typical highly expansive soils which occupy about 20% of total area of country. They exhibit a wide range of shrinkage and swelling characteristics and give rise to cracks of 7–15 cms wide and up to 3 m deep on drying. These soils cover about 0.8×10^6 km^2 areas which is more than one-fifth of the country and extend over the states of Maharashtra, Gujarat, Southern part of Utter Pradesh, eastern part of Rajasthan, southern and western part of Madhya Pradesh, and few parts of Andhra Pradesh and Chennai. The usage of biodegradable natural materials is becoming popular in rehabilitation of areas damaged either by natural or industrial causes, especially in the light of growing awareness of sustainable development throughout the world (Sharma and Ravindranathan 2012).

The composition of the black cotton soil considerably varies with different depth horizons. The clay fraction of the black cotton soil is very rich in silica 60% and iron 15% with only aluminium about 25%. Properties of BC soils, like Liquid limit, Plastic limit, Plasticity index and Shrinkage limit, are very high and CBR is very less. So the soil behavior is objectionable. There are many methods of soil stabilization, like lime stabilization, cement stabilization, mechanical stabilization, chemical stabilization etc., for improvement of expensive soils. But they do not pulverize in actual field condition. Now a days, Coir geotextile-reinforced soil may be used as a soil improvement technique, with respect to embankment, subgrade/subbase, and other problems. Soil bio-engineering with coir geotextiles finds effective application in many field situations. Johnson and Gopinath (2016) studied on swell Behaviour of Expansive Clays Reinforced with Saw Dust, Coir Pith & Marble Dust. Jose and A T T (2017) studied on effect of coir pith and quarry dust on geotechnical properties of expansive soil. Improvement from Coir geotextile implies an increase in the bearing capacity of roadway sections as well as providing separation and drainage between the subgrade and subbase layers.

4 Conclusions

1. The bearing capacity improved by providing coir geotextiles as reinforcement. It is reported an increase in bearing capacity by 1.83 times for reinforced pavement compared to unreinforced pavement.
2. The soaked CBR value also improved when coir geotextiles are used.
3. Improvement from Coir geotextile implies an increase in the bearing capacity of roadway sections as well as providing separation and drainage between the subgrade and subbase layers.
4. Coir geotextiles are found to last for four to six years within the soil environment depending on the physical and chemical properties of the soil.

References

Azeez, A., Evangeline, S., Sayida, M.: Effect of natural geotextiles on unpaved and paved road models- a comparative study. In: Indian Geotechnical Conference (2015)

Sasi, Anu J., Aparna, Sai J.: Subgrade soil improvement using coir geotextile and geo cells. IJERT **5**(8), 1–5 (2017)

Babu, K.K., Beena, K.S., Raji, A.K.: Design of Coir Geotextile reinforced Road using IRC method. Highway Res. J. Special Issue (2008)

Barksdale, R.D., Brown, S.F. Chan, F.: Potential Benefits of Geosynthetics in Flexible Pavement Systems, National Cooperative Highway Research Program Report No. 315, Transportation Research Board, Washington DC (1989). 56p

Baruah, U.K., Saikia, B.D., Bora, P.K.: Road construction in assam by using coir mat. In: Indian Geotechnical Conference, GEOtrendz (2010)

Bell, J.R., Steward, J.E.: Construction and observation of fabric retained soil walls. In: Proceedings of the International Conference on the Use of Fabrics in Geotechnics, 20–22 April 1977, vol. 1, pp. 123–128 (1977)

Bergado, D.T., Youwai, S., Hai, C.N., Voottipruex, P.: Interaction of nonwoven needle-punched geotextiles under axisymmetric loading conditions. Geotext. Geomembr. **19**, 299–328 (2001)

Bhosale, S.S., Kambale, B.R.: Laboratory study on evaluation of membrane effect of geotextile in unpaved road. In: The 12th International Conference of International Association for Computer Methods and Advances in Geomechanics, Goa, India, pp. 4385–4391 (2008)

Burd, H.J.: Analysis of membrane action in reinforced unpaved roads. Can. Geotech. J. **32**, 946–956 (1995)

Bhole, C.R., Sunitha, V., Mathew, S.: Effect of coir geotextile as reinforcement on the load settlement characteristics of weak subgrade. In: 6th International Conference on Structural Engineering and Construction Management 2015, Kandy, Sri Lanka, 11–13 December 2015

Chew, S.H., Tan, S.A., Leong, K.W.: Performance of geotextiles stabilized unpaved road systems subjected to pretensioning. In: Geo-Frontiers-2005, Annual ASCE Conference Proceedings, Austin, Taxes, USA, vol. 155, pp. 405–412 (2005)

Das, B.M., Khing, K.H., Shin, E.: Stabilization of Weak Clay with Strong Sand and Geogrid at Sand-Clay Interface, Transportation Research Board, 1611, National Research Council, National Research Council, Washington, D.C., pp. 55–62 (1998)

Dhule, S.B., Valunjkar, S.S., Sarkate, S.D., Korrane, S.S.: Improvement of flexible pavement with use of geogrid. Electron. J. Geotech. Eng. **16**, 269–279 (2011)

Duncan-Williams, E., Attoh-Okine, N.O.: Effect of geogrid in granular base strength – an experimental investigation. Constr. Build. Mater. **22**, 2180–2184 (2008)

Elvidge, C.B., Raymond, G.P.: Laboratory survivability of nonwoven geotextiles on open-graded crushed aggregate. Geosynthet. Int. **6**(2), 93–117 (1999)

Erickson, H.L. Drescher, A.: Bearing capacity of circular footings. J. Geotech. Geoenv. Eng. **128**(1), 38–43 (2002)

Fannin, R.J., Sigurdsson, O.: Field observations on stabilization of unpaved roads with geotextiles. J. Geotech. Eng. ASCE **26**(7), 544–553 (1996)

Giroud, J.P., Noiray, L.: Geotextile reinforced unpaved road design. J. Geotech. Eng. ASCE **107**(9), 1233–1254 (1981)

Giroud, J.P., Han, J.: Design method for geogrid-reinforced unpaved roads. I: Development of de-sign method. J. Geotech. Geoenviron. Eng. **130**(8), 775–786 (2004)

Giroud, J.P., Ah-Line, C., Bonaparte, R.: Design of unpaved roads and trafficked areas with geo-grids. In: Polymer Grid Reinforcement. Thomas Telford Limited, London, pp 116–127 (1985)

Hu, Y.C., Zhang, Y.M.: Analysis of load-settlement relationship for unpaved road reinforced with geogrid. In: First International Symposium on Geotechnical Safety and Risk, Tongji University, Shanghai, China, pp. 609–615 (2007)

IRC: 81–1997 Guidelines for Strengthening of Flexible Road Pavements using Benkelman Beam Deflection technique

IRC : SP 72 – 2007 Guidelines for the design of flexible pavements for Low volume Rural Roads

IS: 2720(Part 16) – 1979. Methods of Test for Soils: Determination of California Bearing Ratio. [12] IS 2720 (Part 5)-1985, Methods of Test for Soils: Determination of Atterberg Limits. Performance study on Coir Geotextiles in pavements having soft soil subgrade | 45

IS 2720 (Part 8) – 1983. Methods of Test for Soils: Determination of Water Content – Dry Density Relation using Heavy Compaction

IS: 15868 (Parts 1 to 6) – 2010. Methods of test natural fibre geotextiles (Jute geotextile and coir Bhoovastra)

Jose, S., A T T: Effect of coir pith and quarry dust on geotechnical properties of expansive soil. In: International Conference on Geotechniques for Infrastructure Projects (2017)

Johnson, S., Gopinath, B.: A study on the swell behaviour of expansive clays reinforced with saw dust. Coir Pith Marble Dust. IJERT 5(9), 565–570 (2016)

Kazimierowiicz-Frankowska, K.: Influence of geosynthetic reinforcement on the load-settlement characteristics of two-layer subgarde. Geotext. Geomembr. 25, 366–376 (2007)

Kumar, P.S., Rajkumar, R.: Effect of geotextile on CBR strength of unpaved road with soft subgrade. In: EJGE, vol. 17, pp. 1355–1363 (2012)

Love, J.P., Burd, H.J., Milligan, G.W.E., Houlsby, G.T.: Analytical and model studies of reinforcement of a layer of granular fill on a soft clay subgrade. Can. Geotech. J. 24, 611–622 (1987)

Ling, H.I. Liu, Z.: Performance of Geosynthetic-Reinforced Asphalt Pavements. 127(2), 177–184 (2001)

Maxwell, S., Kim, W., Edil, T.B., Benson, C.H.: Effectiveness of geosynthetics in stabilizing soft subgrades. Report to the Wisconsin Department of Transportation (2005)

Mehndiratta, H.C., Chandra, S., Singh, V.: Correlations amongst strength parameters of soil reinforced with geotextiles. Highway Research Bulletin 49 (Indian Roads Congress, New Delhi), pp 13–24 (1993)

Michael, M., Vinod, P.: California bearing ratio of coir geotextile reinforced subgrade. In: 10th National Conference on Technological Trends, College of Engineering, Trivandram, India, pp. 63–67 (2009)

Naeini, S.A., Mirzakhanlari, M.: The effect of geotextile and grading on the bearing ratio of granular soils. Electron. J. Geotech. Eng. vol. 13, Bundle J, Paper 0891 (2008)

Naeini, S.A., Moayed, R.Z.: Effect of plasticity index and reinforcement on the CBR value of soft clay. Int. J. Civ. Eng. 7(2), 124–130 (2009)

Nair, A.M., Latha, G.M.: Bearing resistance of geosynthetic reinforced soil-aggregate systems. In: Proceedings of International Conference on Advances in Materials, Mechanics and Management, College of Engineering, Trivandram, India, vol. I, pp. 457–463 (2010)

Nair, A.M., Latha, G.M.: Bearing resistance of reinforced soil aggregate systems. Ground Improvement 164(2), 83–95 (2011)

Nithin, S., Sayida, M.K., Sheela Evangeline, Y.: Experimental investigation on coir geotextile reinforced subgrade. In: Proceedings of Indian Geotechnical Conference, Delhi (2012)

Oloo, S.Y., Fredlund, D.G., Gan, J.K.-M.: Bearing capacity of unpaved road. Can. Geotech. J. 34, 398–407 (1997)

Raji, A.K., Amruthalekshmi, G.R., Anuj, K., Peter, M., Mohamed, S.: Study of rut behaviour of coir reinforced black cotton soil using wheel tracking apparatus. In: Proceedings of Indian Geotechnical Conference, Kochi (2011)

Ranadive, M.S., Jadhav, N.N.: Improvement in bearing capacity of soil by geotextiles-an experimental approach. In: 5th International Conference on Ground Improvement Techniques, Malaysia, G104 (2004)

Rao, G.V., Balan, K.: Coir Geotextiles- Emerging Trends. The Kerala state coir corporation Limited (publishers) Alappuzha, Kerala (2000)

Raymond, G., Ismail, I.: The effect of geogrid reinforcement on unbound aggregates. Geotext. Geomembr. **21**, 355–380 (2003)

Retzlaff, J., Turezynski, U., Schwerdt, S.: The effect of geogrids under unbound sub-base layers. In: Proceedings of the 8th International Conference on Geosynthetics, Yokohama, Japan, vol. 3, pp. 825–830 (2006)

Sharma, U.S., Ravindranathan, A.D.: Application of coir geotextiles in the construction of roads on agrarian soils. In: National Workshop on Non-Conventional Material Technologies, NRRDA, New Delhi (2012)

Som, N., Sahu, R.B.: Bearing capacity of a geotextile-reinforced unpaved road as a function of deformation - a model study. Geosynth. Int. **6**(1), 1–17 (1999)

Srivastava, R.K., Jalota, A.V., Singh, R.: Model studies on geotextile reinforced pavements. Indian Highways **23**(9), 31–39 (1995)

Subaida, E.A., Chandrakaran, S., Sankar, N.: Experimental investigations on tensile and pullout behaviour of woven coir geotextiles. Geotext. Geomembr. **26**(5), 384–392 (2008)

Subaida, E.A., Chandrakaran, S., Sankar, N.: Laboratory performance of unpaved roads reinforced woven coir geotextiles. Geotext. Geomembr. **27**, 204–210 (2009)

Venkatappa Rao, G., Dutta, R.K., Ujwala, D.: Strength characteristics of sand reinforced with coir fibers and coir geotextiles. In: EJGE, vol.10 (2005)

Watts, G.R.A., Blackman, D.I.: The performance of reinforced unpaved sub-bases subjected to trafficking. In: Third European Geosynthetics Conference, German Geotechnical Society and Zentrum Geotechnick, pp. 261–266 (2004)

Cantilever Segmental Retaining Walls

Chip Fuller[(⊠)]

Abaco Technologies, Inc., Dubai, UAE
chipfuller@mindspring.com

Abstract. A unique entry into the technology of segmental retaining walls is becoming popular in the U.S. and several other countries, including Egypt where it has been approved and used extensively for highway retaining walls. Using a well-proven concept of cantilever-enhanced stability, the Cantilever Segmental Retaining Wall (CSRW) system creates an extremely stable facing system, combined with mechanically-stabilized earth (MSE) technology to offer new options for engineers to consider. Of particular interest is the ability of the CSRW system, comprised of a two-part concrete facing system combined with geogrid-reinforced soil, to be wet-cast produced on the jobsite in lightweight reusable molds. This creates a new reduced-cost paradigm while maintaining acceptable engineering factors of safety in accordance with international standards. This paper proposes to present the core science behind this innovation, as well as pertinent construction and benefit details for a full 360° view of how the CSRW system could become a displacement retaining wall technology in many geographic areas.

1 Introduction

Retaining walls have been part of human history for thousands of years. The need to support steep hillsides, embankments, and human-made earthen structures was a key civil engineering challenge as we developed into organized societies.

For the great majority of our history, retaining walls were relatively simple gravity structures. Early builders knew that mounds of rock would withstand the lateral earth pressures if the thickness of the mound was sufficient (Figs. 1 and 2). The Greeks and Romans developed simple mathematical methods of determining how thick these stone walls had to be for sufficient stability.

2 Types of Engineered Walls

2.1 Concrete Gravity Walls

With the ancient development of concrete, it was sometimes easier and more economical to build a gravity mass out of primitive concrete. This required the use of forming systems and more skilled labor, but it was used for centuries to create stable walls, particularly in areas with poor sources of rock (Fig. 3).

M. Meguid et al. (Eds.): GeoMEast 2018, SUCI, pp. 165–186, 2019.
https://doi.org/10.1007/978-3-030-01944-0_13

Fig. 1. Ancient Roman walls

Fig. 2. Concrete gravity wall

2.2   Cantilever Walls

The era of modern retaining walls using engineered concrete started with the development of steel-reinforced concrete in the early 1920s. The steel reinforcement gave
the concrete enhanced tensile strength to counteract the lateral earth pressures and
induced bending moments in the concrete, allowing thinner concrete cross-sections to

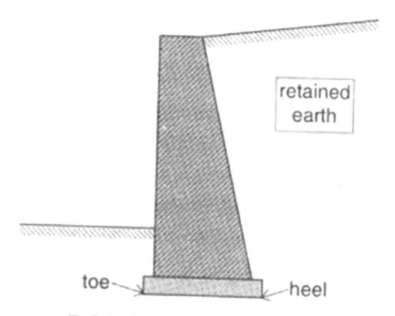

Fig. 3. Simple gravity wall made of rock or concrete

be used. But the major development that significantly reduced the cost and time of construction was the cantilever retaining wall. Cantilevered walls are made from a relatively thin stem of steel-reinforced, cast-in-place concrete, often in the structural footing, converting horizontal pressures from behind the wall to vertical pressures on the ground below. Sometimes cantilevered walls include a counterfort on the back, to improve their stability against high loads (Fig. 4). This type of wall used much less concrete than a traditional gravity wall. Cantilevered walls resist lateral pressures by friction at the base of the wall, increasing the tendency of the soil to resist lateral movement.

Previously, these retaining walls were commonly cast-in-place concrete cantilevers. They were rigid and monolithic in design and were a significant project expense. The base of these types of structures supported the soil above it and in turn restrained the horizontal earth pressure on the vertical wall section thru its interaction with the foundation soils and the passive resistance of the soil.

2.3   Mechanically Stabilized Earth (MSE) Walls

Starting in 1971, The Reinforced Earth Company commercialized the concept of Mechanically Stabilized Earth (MSE) structures. Using relatively thin precast concrete panels for the facing, overall stability of the wall was realized by extending steel strips from the back of the panels into the "reinforced soil" behind the walls. This created a stabilized soil mass that essentially behaved like an ancient stone gravity wall. The quantity, strength, spacing, and embedment depth (length) of the steel strips was determined by engineering calculations that were developed and standardized. Since

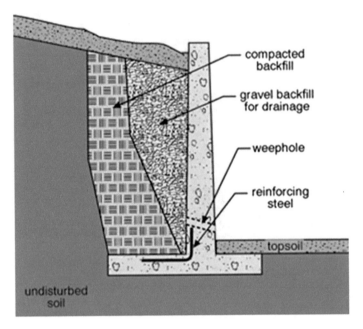

Fig. 4. Cantilever retaining wall

1971 this has been a proven and popular retaining wall technique and paved the way for additional innovation (Figs. 5 and 6).

2.4   Segmental Retaining Walls (SRW)

The first polymeric geogrids were brought from the UK into the United States in 1982. As with RECO's steel strips, these geogrids were designed to reinforce the soil in the reinforced zone to such an extent that the entire reinforced earthen structure became a stable gravity mass. Unlike steel strips, the polymeric geogrids did not corrode and thus could be used with most in-situ soils in the reinforced zone instead of very select (and expensive) backfill soils/aggregates. This led to the rapid development of segmental retaining walls, which combined relatively small, interlocking concrete facing blocks with geosynthetic-reinforced soil. SRW facing blocks (a few are shown in Fig. 7) provided an extremely durable and aesthetic concrete facing and allowed the entire structure to be slightly flexible, meaning it could settle along with the small post-consolidation of the soil without building long-term stress in the concrete facing. This produced a classic displacement technology, replacing standard bulky and expensive reinforced concrete walls (most of them cast-in-place cantilevered) that had been the state of practice for many years. Using MSE design methods, SRWs became an accepted method of building retaining walls in many developed countries in only 10 years. A wide range of low-cost, attractive concrete facing blocks were introduced, concurrent with ASTM testing protocol and engineering design standards. SRW design and construction methods were researched by the U.S. Federal Highway

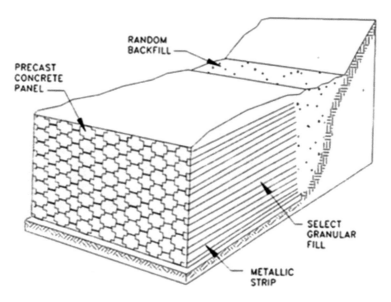

Fig. 5. RECO wall

Fig. 6. MSE wall under construction

Administration, AASHTO, and the US-based National Concrete Masonry Association, as well as various European agencies and associations, all of which helped propel this improved technology forward (Figs. 8 and 9).

For the past 20 years, SRW retaining walls, using concrete block facing systems, have been used in virtually every major country in the world due to their relatively low cost, high factors of safety, and ease of construction. The question now is: What's next?

Fig. 7. Various SRW blocks (NCMA)

Fig. 8. Example of segmental retaining walls (Courtesy of Rockwood/Anchor Retaining Walls)

3   Important Developments

3.1   The Introduction of Precast Concrete Cantilever Retaining Walls

An early innovator in precast (i.e., *not* poured-in-place) concrete retaining walls was Mr. David Ash, who began his work in the modularization of the cantilever retaining concept in the 1990s with a U.S. company called Stresswall. Stresswall utilized large, reinforced concrete cantilever brackets weighing several tons combined with equally large concrete panels supported between the cantilever brackets (see Figs. 10, 11 and 12).

Fig. 9. Example of segmental retaining walls (Courtesy of Versa-Lok Retaining Walls Co.)

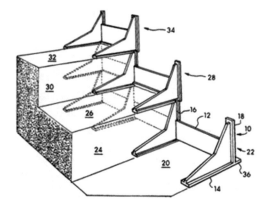

Fig. 10. From original Stresswall patent

StressWall was the first company to modularize the cantilever retaining wall both horizontally and vertically, and to make it available for off-site production, trucking the large modules to the jobsite and lifting them into place with a crane. The vertical consolidation of the compacted backfill soils was accommodated by the layers of cantilevers not needing support from the elements below (not stacking on top of one another); instead, they were supported by the footing structure that projected into the backfill. This system was very effective and used in the U.S. for many highway retaining walls, but the units were too heavy. As concrete block facings with geogrids achieved acceptance in the industry, Stresswall became financially non-competitive and exited the market. However, it served the purpose of conceptualizing and creating precast concrete facing modules with a cantilever component to resist overturning.

Fig. 11. Stresswall project example

Fig. 12. Stresswall project example

3.2 Combining SRW and Cantilever Technology – A Breakthrough

As the development of soil reinforcements progressed, Mr. Ash invented a new type of retaining wall using four important, fundamental concepts:

1. MSE theory of reinforcing in-situ soil with polymeric geogrids to create a stable gravity structure
2. Using a cantilever component attached to a concrete facing panel to stabilize and add structural unity to the entire wall

3. Using durable, precast concrete units for the wall facing which can be produced in reusable molds on or near the jobsite
4. Designing the system to optimize the largest hand-placed (no machinery required) components possible for rapid installation using unskilled labor

This led to the innovation of a new category of retaining walls called Cantilever Segmental Retaining Walls (CSRW). The system created in this invention has the brand name LOCK+LOAD®, which consists of a two-part, precast, reinforced concrete facing and counterfort unit (Fig. 13). When combined with geogrids to reinforce the soil behind the facing system, a hybrid cantilever MSE wall is created.

Fig. 13. CSRW system components and dimensions (LOCK+LOAD®)

As explained by Pimentel and Pereira (2012), the basic concept of the system consists of filling the modules, which measure 660 mm D × 406 mm H × 813 mm W, with soil, generating a region for anchoring the geogrid reinforcements and promoting the confinement of the counterfort that ensures the locking of the face panel. This allows it to withstand the high lateral loads of compacting the reinforced soil directly behind the wall with heavy equipment. The essential components of this CSRW system are the lightweight (40 kgs) concrete facing panels that "lock" into a reinforced concrete counterfort, which measures 660 mm long and weighs approximately 20 kg (Figs. 14 and 15). The locking mechanism (Fig. 14) is a loop cast into the back of the facing panel, made of either stainless or galvanized steel, which tightly engages with a notch at the head of the counterfort (International Code Council – Evaluation Service 2002).

Once the backfill soil/aggregate is placed in designed lifts and well compacted, the modular system is "loaded" and becomes a structural block that fully engages the geosynthetically-reinforced soil with the concrete components (Fig. 16).

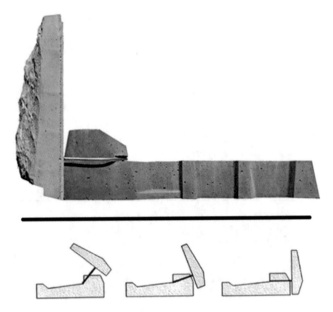

Fig. 14. Loop connection (top) and locking mechanism (bottom)

Fig. 15. Precast notch for the locking of the loop

A strong advantage of the system is that the concrete modules do not rest directly on top of each other and are thus independently stable. The granular material, when

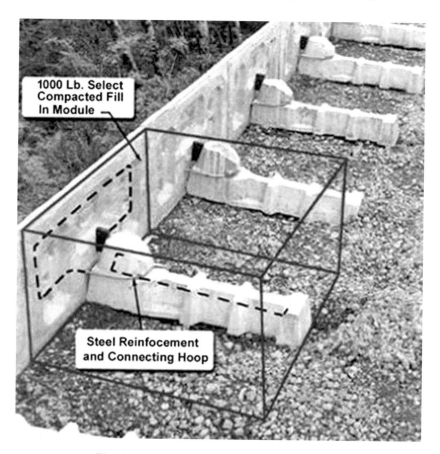

Fig. 16. Zone of influence due to cantilever effect

placed on the counterfort and behind the panel, promotes the effective locking of the parts, forming a rigid and stable assembly. This rigid block forms the foundation for the subsequent row of modules, providing adequate support without the risk of surcharging each concrete facing panel with the other panels above it (Fig. 17). This feature allows this system to create a wall with a face that is nearly vertical.

The advantages of the CSRW versus traditional SRW walls are significant:

1. The CSRW panels have larger face dimensions than the segmental blocks: one LOCK+LOAD® facing panel is equal to 3.5 SRW blocks, while maintaining equivalent unit weights. This creates better labor productivity and faster installations.
2. The facing panels are cast face-down in reusable molds which can have various patterns molded into the bottom of the mold. This affords these walls with a natural rock-like appearance for better aesthetics.

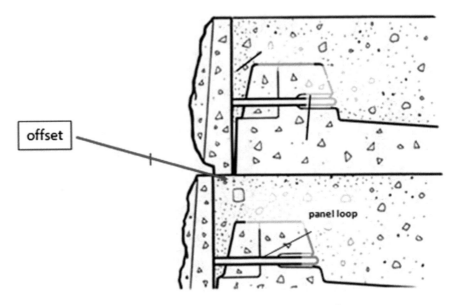

Fig. 17. Note the offset between facing panels

3. Both the panels and counterforts are substantially reinforced with both steel and polymeric fibers, creating concrete strengths of approximately 4,500 psi vs. conventional SRW's 2,500 psi.
4. The CSRW concrete units, because they are not made by large automated block machines in a factory, can be produced on or near the jobsite in a customized, simple production facility using durable, lightweight polymeric molds.
5. Due to the cantilever effect of the counterfort, large compaction equipment can operate directly behind the facing panel without the risk of the panel tipping forward. This total soil compaction in the near-face area is critical (Fig. 18).

Using classic cantilever technology, this system reduces the amount of concrete needed in the facing system by 50% compared to standard SRW block systems and 70% as compared to large, wet-cast concrete, modular systems.

4   Cantilevered SRW System Details

4.1   Production of the Components

The LOCK+LOAD® system is wet-cast with fiber-reinforced concrete in rigid plastic molds that can withstand over 800 production cycles. Each mold is inexpensive and weighs only 30–40 kg. The facing panels, which measure 405 × 810 mm, are cast face down in the mold and easily stripped out of the mold after 24 h. One production facility in Portland, Oregon, USA consistently has a production output of 1,000 units per day (320 m^2 of wall face). In Brazil, a producer is using a more basic production setup. With 4 laborers and one supervisor on the CSRW production team, the

Fig. 18. Showing high surcharge load directly behind the facing unit with no movement or distress

production is 40 m^2 of wall face per day. The stacking patterns of the facing panels and counterforts on pallets make for easy handling. These techniques are well-established and maximize efficiency. The use of unskilled labor, reduction in transportation costs, and simplified production combine to make this a cost-saving proposition, particularly in developing countries, while maintaining standards of quality in all aspects of the final product. The concrete in the panels and counterfort should be tested at an independent laboratory on a regular statistical basis for compressive strength and absorption (Fig. 19).

4.2   Installation

CSRW systems are installed in a similar procedure to standard SRW block walls, with one important difference. To begin, a bottom row of facing units is carefully placed upon a prepared foundation in accordance with the elevations shown on the plans. Each unit is tamped into place and leveled. Then, one or more layers (lifts) of backfill soil are placed and compacted to 95% Procter. With SRW blocks, the next row of blocks is placed directly on top of the row of blocks below it, and the backfilling and compaction of the soil is repeated. (It is highly recommended that the density of the compacted soil

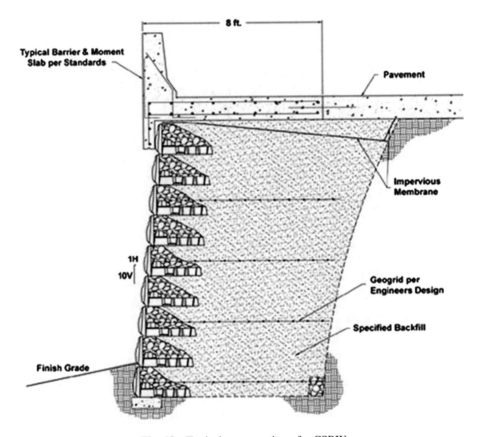

Fig. 19. Typical cross section of a CSRW

be checked at regular intervals.) However, the CSRW system does not place upper rows of concrete units directly on the units below. Instead, the units are placed on the *soil*, slightly offset from the units below. This allows them to act independently and not transfer surcharge stresses to the units below, instead relying on the inherent stability of the attached counterfort. With either system, the layers of geogrid soil reinforcement are placed at elevations shown on the plans.

5   Case Study: Warehouse Project in Brazil

As an example of a typical project, the following will explain the installation of a CSRW wall in Brazil.

5.1   Project Background

The work on this project was intended to increase the useful space of an industrial warehouse complex located near São Paulo, Brazil. Figure 20 shows the site of the facility where the wall was built. The work consisted of constructing a retaining wall to extend the parking area in front of the existing building on a slope.

Fig. 20. Initial view of the area where the wall was executed

For the design of the wall, the external and internal stability criteria were used. For the external stability criteria, safety factors were set at 1.5 for sliding, 2.0 for over-turning, 3.0 for foundation bearing capacity, and 1.5 for global failure.

For internal stability criteria, the Mitchell & Villet method (1987) was used. In this method, it is assumed that the soil reaches the failure state of Rankine. That is, the hypothetical surface of failure is oriented along a plane forming an angle of $(45° + Ø/2)$ with the horizontal. This results in a linearly increasing distribution of forces with depth in the reinforced mass. For sizing the geogrid reinforcement, geogrid reduction factors were used for installation damage, creep, chemical degradation, and biological degradation.

For the CSRW wall, geogrids were adopted with three different tensile strengths. Geogrids with nominal resistance of 55 kN/m were used in the upper portion of the wall and 70 kN/m were used in the middle portion. In the lower portion, where the largest loads were expected, the geogrids provided a nominal resistance of 90 kN/m.

5.2   Production of the Components

In this project near Sao Paulo, due to the large number of panels used (approximately 4,000 units at approximately 3 units per 1 m^2 of face area), a temporary production plant (Fig. 21) was installed on site to produce the facing panel and counterfort components. Form molds were filled with fiber-reinforced concrete, left to cure for 24 h, and removed from the molds. The cast components were stored for curing and inventory prior to installation. For this application, the modules were specified in concrete with a strength of 35 MPa with a 3% additional volume of polypropylene reinforcement fibers.

Fig. 21. Temporary production yard in Brazil near the project

5.3   Installation Sequence

The engineering design required the bottom of the wall to be embedded 1 m beneath the finished grade. Thus, the first step of the installation was the excavation for the base of the wall and improving the foundation with a layer of gravel (Fig. 22).

Fig. 22. View of the excavation for embedment (Geo Soluções 2013)

For the placement of the panels and counterforts, the panels were positioned face down along the wall face. After this, the counterforts were locked into the panels by the connecting loops. The units were manually rotated to have the cantilever resting on the ground while securing the loop of the panel to verticalise and lock the face panel (Fig. 23). With the panels mounted and aligned side by side, their verticality was verified by means of leveling adjustments of the gravel layer under the couterfort, with the panels installed along the wall alignment (Fig. 24).

Fig. 23. Locking the counterfort to the facing panel

After the installation of the facing panel, counterfort, and any required geogrid, backfill was spread and compacted using a 400 kg vibrating plate (Figs. 25 and 26).

In the "retained zone" a cohesive local soil with 60% fines, specific weight of 17 kN/m^3, and friction angle of 33° was used. This fill was spread over the geogrid reinforcement layer and compacted in 20 cm thick layers.

In some corner sections there was a need to cut some panels so as to provide adequate fit, as shown in Fig. 27.

With the use of the LOCK+LOAD® system, the installation speed of the reinforced soil was higher than that normally observed with other retaining wall facing systems. The work was executed in 79 consecutive days, with 44 days actually worked, discounting weekends, mobilization and demobilization of teams, and non-productive days (rainy days). With the total face area equal to 1,610 m^2, the speed of installation including the earthwork was equal to approximately 37 m^2 of face area/day, which corresponds to the installation of 117 panels per day. This speed of execution was achieved with a team of 5 laborers. Adequate inspection by an experienced engineer was performed throughout the project.

Fig. 24. Alignment of panels along the wall (Geo Soluções 2013)

Fig. 25. Spreading the soil on the geogrid for later compaction

Fig. 26. Compaction of the backfill material in the section of the counterfort (Geo Soluções 2013)

Fig. 27. Cut corner panels

5.4   Completed Project Observations

This case study presents a CSRW project constructed in Brazil in which the owner needed a very cost-effective wall, built to international-quality standards, and with an attractive face for the public. Also, the project schedule was very tight. Careful coordination was needed with the earthworks contractor and other subcontractors.

The main observations about this case study are:

a. The total installed cost of the project in this location was approximately 20% less expensive than any other concrete-faced system available in this location.
b. The wall was built on time and under budget.
c. The majority of the installation crew were unskilled laborers who were supervised by an experienced foreman. No heavy equipment was required except for earthwork.
d. This 5-year-old project shows no signs of movement, cracking, or distress.

It is significant that the panels contributed to an aesthetically pleasing retaining wall, generating a greater added value for the warehouse complex (Fig. 28).

Fig. 28. General view of the finished wall

6 Conclusion

This paper has provided an introduction to a new type of retaining wall system which may have significant applications in many countries, particularly areas with limited access to automated concrete block factories, plenty of semi and unskilled labor, and a demand for more cost-effective concrete-faced retaining walls. Like all new innovations, there is a learning curve to the manufacturing and installation processes of a CSRW system; but a production license does not require a large initial financial outlay, the business is easily scalable as the demand increases, and this technology has a proven record of cost savings, technical back-up, and successful projects.

On a local basis it should be noted that there are many successful LOCK+LOAD® projects completed in the Cairo area (Figs. 29 and 30).

Fig. 29. A finished CSRW project in Egypt

OGDEN, UT USA   HIGHLAND VALLEY, B.C. CANADA   GOLDEN VALLEY ROAD, SANTA CLARITA, CA USA

Fig. 30. Typical CSRW projects

References

Geo Soluções: Internal Work Control. Geo Soluções, São Paulo (2013)

International Code Council – Evaluation Service: Engineering Report 5893 (2002). http://www. icc-es.org/Reports/pdf_files/5893.pdf. Accessed 31 May 2018

Mitchell, J.K., Villet, W.C.B.: Reinforcement of earth slopes and embankments. National highway research program report, no 290 (1987)

Pimentel e Pereira: GRS walls with facings in pre-casted concrete, Revista Techne, abril de 2012, 7 p

Pre-stressed Segmental Retaining Walls (PSRWs)

Reza Hassanli[1], Md Rajibul Karim[1(✉)], Md Mizanur Rahman[1],
Arman Kamalzadeh[2], Julie Mills[1], and Mehdi Javadi[1]

[1] School of Natural and Built Environments, University of South Australia,
Mawson Lakes, SA, Australia
Rajibul.Karim@unisa.edu.au
[2] Faculty of Engineering, University of Auckland, Auckland, New Zealand

Abstract. This paper introduces an innovative system of retaining walls named "pre-stressed segmental retaining walls (PSRWs)". In this system, interlocking blocks are assembled together with dry joints (mortarless) and the integrity of the wall is maintained by pre-stressing forces. The pro-posed system has a collection of advantages over the conventional systems for construction of cantilever retaining walls or mechanically stabilized earth walls. In particular precast concrete/masonry segments can be incorporated which reduces the construction time and cost for cantilever type structures and if combined with mechanically stabilized earth wall system, it can reduce the number of layers of reinforcement and add flexibility to the design. These walls will be suitable for both water front and soil retention purposes.

1 Introduction

To overcome the weak tensile capacity of concrete/masonry, steel reinforcement has been used in concrete/masonry structures. However, the main drawback of this is that the steel does not contribute to the strength until tensile cracks occur. To overcome this, post-tensioning (PT) of structural members has been introduced. PT offers engineers the possibility of actively introducing desired stresses to different structural members to enhance their strength, cracking behavior and ductility (Ganz 2003). PT can be inte-grated with precast concrete and masonry in modular construction to provide fast and cost-effective solutions to many problems such as segmental walls/columns where PT effectively connects the precast units and provides structural integrity to the system. PT is often most beneficial in situations where the lateral load (e.g. wind, earthquake or lateral earth/water pressure), is high compared to the axial load (Bean Popehn et al. 2007). In recent years, a number of studies have been carried out to investigate the behavior of segmental self-centering concrete columns (Chang et al. 2002; Chou and Chen 2006; Chou and Hsu 2008; ElGawady and Sha'lan 2011; Hassanli et al. 2017; Hewes and Priestley 2002; Kwan and Billington 2003; Marriott et al. 2009; Ou et al. 2007). The use of precast segmental construction for concrete bridges has increased recently due to the demand for shorter construction time and the desire for innovative designs that yield safe, economical and efficient structures (Kim et al. 2010).

© Springer Nature Switzerland AG 2019
M. Meguid et al. (Eds.): GeoMEast 2018, SUCI, pp. 187–196, 2019.
https://doi.org/10.1007/978-3-030-01944-0_14

A very limited number of studies in have been reported that involves introduction of vertical pre-stressing (either bonded or unbonded) to modular retaining wall constructions (cantilever or mechanically stabilized earth walls). All the studies reported in the literature were related to mechanically stabilized earth walls which includes pre-stressing of the vertical wall facing elements. Koh et al. (2013) presented a numerical investigation studying the settlement reduction effect of a back to back reinforced retaining wall with prestressed vertical wall facings. Ahmadi and Bezuijen (2018) reported pre-stressing can increases the rigidity of wall face in a mechanically stabilized wall and results in requiring less geo-reinforcement in comparison to more flexible ones. Tarawneh et al. (2018) reported that 73% of mechanically stabilized earth wall problems (including problems resulted by backfill material leaking from the joints of the facing wall, vegetation growing in the joints of the panels, cracked or deteriorated segments, bowing or bulging wall) can be directly solved using post tensioning. This paper introduces the concept of "Pre-stressed Segmental Retaining Walls (PSRWs)". A PSRW can be a cantilever retaining wall or mechanically stabilized wall or a combination of both.

2   Retaining Walls

Retaining walls are structures that are constructed to retain soil, water or any other materials unable to stand vertically by themselves because of high lateral stresses. Examples of providing vertical support to retain mass can be found from as early as 4000 BC in Europe (Kerisel 1993). The engineering understanding of the behaviour of retaining structures and their interaction with the retained material started in early 18th century and is still ongoing.

Retaining structures can be of many different types. A broad classification can be, gravity walls, embedded walls and mechanically stabilized walls (MSWs). Each of these groups can be sub-classified into many different types. For example, gabions, crib walling, interlocking block walls, masonry walls, reinforced concrete cantilever walls can all be treated as gravity walls as the stabilizing forces are provided by their self-weight and the weight of the retained materials.

Selection of the type of retaining wall for a particular purpose depends on different factors, such as, function, constructability, appearance, availability, maintenance, re-use, cost and construction time. Different types of walls are often constructed only to certain heights as beyond that they may not be economical choices. For example, reinforced concrete cantilever walls are often constructed to a maximum height of 6 m. For greater heights counterfort walls can be a better choice but are difficult in terms of construction. MSWs are gaining popularity as they can be a suitable and cheaper alternative for walls higher than 5 m. Using mortarless precast units for the construction of MSWs allows installation to proceed rapidly, which results in about 25%–45% cost reduction compared with conventional cast-in-place concrete retaining walls (Bathurst and Simac 1994). Table 1 presents a relative comparison for some of the different types of retaining structures in common use today.

This paper discusses the potential of applying the pre-stressing technique to precast concrete/masonry to remove some of the limitations of two of the most common types

Table 1. Common types of retaining structures and their relative advantages and disadvantages

Wall type	Height* (m)	Advantages/disadvantages
Mass concrete gravity walls	3<	• Suitable for small walls
		• Bulky construction
Gabion walls	2–8	• Very flexible
		• Aesthetically pleasing
		• Shock absorbent
		• Easy construction in remote areas
Crib walls	6–9	• Can tolerate large movements
Interlocking blocks	3<	• Aesthetically pleasing
		• Suitable for small heights
Masonry wall	3<	• Suitable for small walls
		• Double-skinned reinforced and grouted cavity walls are suitable for greater retained heights
Reinforced concrete cantilever walls	6<	• Simple form of L or inverted T
		• Suitable for low to medium height walls
		• For greater heights counterforts or buttresses are needed
Counterfort walls	10–12	• Complicated to build because of the counterforts
		• Can be constructed to greater heights than cantilever walls
Anchored sheet pile walls	5–20	• Can follow complex plan shapes with ease
		• Cause minimal soil displacement during driving
		• Speed of installation and extraction leads to a high degree of sustainability
		• Can be expensive as a permanent solution
		• Traditional methods of installation can be noisy
		• Wall depth is limited by section size, loads, and standard stock lengths
Mechanically stabilized walls	3–10	• Very flexible
		• Retaining structure is made simultaneously with the filling
		• Suited for the construction of highway embankments on steep sidelong ground, abutments and wing walls of bridges
		• Can tolerate large settlements and differential rotations
		• More economical at heights over 5 m

*Construction heights of the walls are limited to the values provided in the table mainly because of economy. Higher walls exist but are uncommon.

of walls that are currently in use i.e. reinforced concrete cantilever walls and MSWs. As explored in the paper, this new type of wall will promote faster construction and a more flexible design approach.

One of the drawbacks of the conventional reinforced concrete walls is their long construction time. Formwork needs to be in place along with the reinforcement and curing needs to be done before the structure can go into service. These often add to the costs, along with other inconveniences in the surrounding area related to the construction. Use of precast units along with post-tensioning have the potential to shorten the construction time significantly in these cases. Among other advantages are a higher level of quality control, flexibility in providing more efficient/complicated shapes of sections, and that pre-stressing of concrete will help to reduce the volume of concrete needed for a job. This will potentially lead to significant economic benefit and a much smaller carbon footprint for the structure. It will also be possible to construct the walls to greater heights economically than was possible using conventional reinforced concrete.

One of the most common types of retaining walls, MSW, consists of dry-stacked modular concrete units connected to the backfill soil using metallic or polymeric reinforcements (geotextile or geogrid). The strength and stability of MSWs come from the frictional interaction between the backfill soil and the reinforcements resulting in a bond that creates a composite structure. These structures can be prone to local failure leading to a total collapse of the structure. Use of post tensioning in the block face of a MSW can potentially eliminate many of the local failure scenarios.

3 Proposed System

3.1 Pre-stressed Segmental Retaining Wall

The use of precast concrete has significantly increased due to its advantages over cast-in-site concrete, such as high economic benefits, less environmental effects, good quality control and time saving. Precast concrete has been used in many different types of structures. Similar to segmental post-tensioned column members, retaining walls can be constructed out of precast concrete/masonry units where the units can be clamped together using pre-stressing. Some of the potential benefits of the proposed PSRW system can be,

- Suitability for a wide range of heights
- Desirable saving on construction time and cost
- Use of precast concrete gives better quality control and higher strength of concrete
- Economic and environmental benefit due to cross sectional decrease
- Possibility to shape the precast concrete sections to forms that is difficult or even impossible in in-situ casting.
- Utilizing the full strength of the concrete section (uncracked section) because of pre-stressing.
- Applicability for retaining soils as well as water front structures.
- Low repair cost

Three types of PSRWs are proposed here and the potentials, challenges and the future studies on this area are explored.

3.1.1   Pre-stressing in Traditional MSWs

In general the failure modes in a MSW system can be grouped into three categories, i.e. external, internal and overall slip instability. Figure 1a–c shows different forms of external instability for a MSW. Base sliding may occur when the resistance against lateral movement is lower compared to the lateral forces acting on the wall. Over-turning may occur when the overturning moment due to the lateral forces acting on the wall about its toe is greater than its resisting moment, which comes from its self-weight and any other permanent vertical forces acting on the wall. Sometimes the foundation soil may not have enough strength to support the retaining structure and lead to a bearing capacity failure. To function, any retaining structure should be safe against these types of failures. Figure 2 shows different forms of internal instability (also known as local failure) which mostly occur due to relative movements of the wall segments (facing units) and the connection failure. Local failure in MSWs includes: Internal sliding, shear failure, local overturning, connection failure, tensile over-stress and pull out.

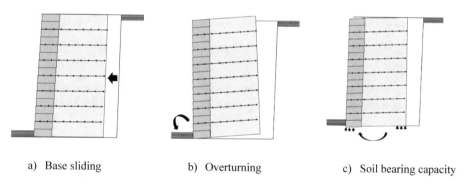

a) Base sliding   b) Overturning   c) Soil bearing capacity

Fig. 1.   Different types of external instability for MSWs

Although it has been shown that MSW is an economical choice compared to other types of retaining wall systems, it has its own limitations. There are many situations where reinforcing the soil is not very effective or not effective at all. For example in soft soil, quick clay, incompatible soils (such as silt or clay), soils where liquefaction might occur, water front structures, or where there is not enough length available for rein-forcement at the back of the wall. To mitigate some of these limitations, post-tensioning (either bonded or unbounded) can be incorporated in MSWs to eliminate many of the local failure modes, as explained below:

Internal Sliding

Internal sliding (Fig. 2a) happens along a reinforcement layer and through the facing unit. Sliding usually occur due to low level of resistance in a part of the wall. By

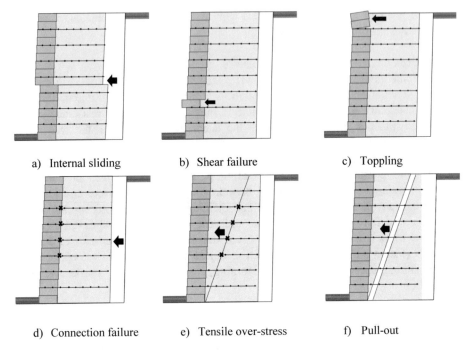

Fig. 2. Different forms of internal instability in MSWs

clamping the wall segments together using post-tensioning, the interaction between the segments increases which resists sliding and relative movements of the blocks.

Shear Failure
Shear failure occurs between the facing units due to low interaction between them to carry shear loads (Fig. 2b). By post-tensioning the wall, the axial load increases the friction between the segments and hence increases the resistance of the wall to shear failure.

Local Overturning (Toppling)
Due to relatively low vertical stress on the facing units located closer to the top of the wall, the friction between those units is comparatively small and might be overcome by soil lateral pressure (especially when large surcharge forces are present near the top of the wall). This will result in local overturning/toppling (Fig. 2c). Inducing axial stress using post-tensioning, increases the interaction between the facing units and resists local overturning of the retaining wall.

Connection Failure
The connection between the wall and the reinforcements might fail if overstressed or not designed/constructed properly (Fig. 2d) and often leads to collapse of the structure. The unit-reinforcement connection capacity controls the spacing and number of the

reinforcing layers. Applying pre-stressing forces can significantly improve the friction to reduce the consequence of connection failures by redistributing the stresses.

Tensile Over-Stress
Tensile overstressing (Fig. 2e) can occur in any layer if improper spacing of reinforcement is used. Local failure of this type often leads to a global failure. However, pre-stressing in the facing elements will help re-distributing the stresses along the height of the wall and thus reduce the possibility of global failure.

Pull Out
Pull out can occur at a local or a global scale in a MSW (Fig. 2f) due to insufficient reinforcement length or improper compaction. Similar to the case of tensile overstress, pre-stressing the facing elements will help redistribute the stresses along the depth of the wall. For a global occurrence, there is often very small relative movement between the facing units. As the main purpose of pre-stressing is to clamp the units together to maintain the integrity of the system, pre-stressing may not be directly effective.

3.1.2 Pre-stressing in Conventional Cantilever Retaining Walls
In the proposed PSRW, precast concrete segments can be used which provides a fast construction solution. Sometimes the main cost of retaining wall construction (especially in cast-in-place reinforced concrete cantilever retaining walls) is related to the low speed of construction. By introducing precast concrete segments and pre-stressing (as shown in Fig. 3 below) the construction can be quicker.

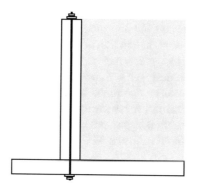

Fig. 3. Proposed variation of cantilever retaining wall incorporating pre-stressing of concrete

3.1.3 Pre-stressing in a Hybrid MSW System
One of the main advantages of the proposed PSRW system is that it can facilitate a combination of different available systems. By having a larger block or footing at the bottom of the wall, as the first facing unit (shown in Fig. 3b and c), a proportion of lateral resistance can be obtained through the cantilever action of the wall. The system will work by combining the mechanism used in cantilever retaining walls (wall toe) and MSWs. This will result in reduction in the reinforcement force to avoid pull-out and

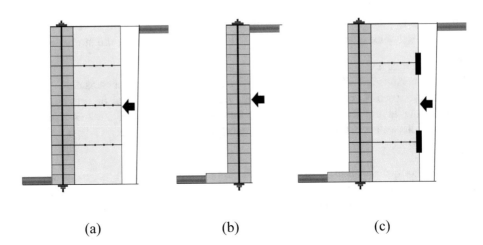

(a)   (b)   (c)

Fig. 4. Possible variations in the proposed PSRW system

tensile overstress failure. This may also reduce the required reinforcement length and may be a useful solution for tight spaces situations. Furthermore, this also opens up the possibility of having fewer reinforcing layers as part of the lateral stress is resisted by the cantilever action (Fig. 4a). In smaller height walls, it may even be possible to construct them without any geo-reinforcements (Fig. 4b). There is also a possibility of using anchor plates or other anchoring systems connected with the reinforcement (Fig. 4c) at the back of the backfill soil to increase the tension resistance force, further decreasing the number of reinforcing layers.

In terms of global failure and stability modes, the advantage of this hybrid system is that both the mechanism used in cantilever retaining walls (wall toe) and MSWs (stabilized soil) contribute to the resistance. The weight of each component in providing global resistance is a designer's choice and depends on the limitations of each design case, however they can be combined to satisfy the stability requirements. This flexibility in design is very helpful for the walls having different limiting conditions. In terms of local failure, as mentioned pre-stressing can potentially prevent all the local mode of failures typically observe in the MSWs.

4   Bonded or Unbonded PSRW?

A post-tensioned wall can be bonded or unbonded. In bonded PSRWs, the ducts through which the PT bars or tendons are passed are grouted using fine aggregate concrete. In unbonded members, the PT ducts are left un-grouted and the PT bars/strands can move freely in the ducts. In an unbonded PT wall the restoring nature of the PT force returns the wall back to its original vertical position and minimizes the residual displacement (self-centering mechanism). This behaviour is specifically favorable for structures which are designed for immediate occupancy performance levels. The unbonded PSRWs present a rocking mechanism, which results in the formation of concentrated plastic deformation at the toe of the wall which can be repaired with minimal cost (Bean Popehn 2007; ElGawady and Sha'lan 2011; Wight 2006).

Depending on the design conditions and limitations, either bonded or unbonded PSRW can be used. For examples, for water front retaining walls, unbonded PSRW can be an appropriate option. Under severe conditions such as flooding, the wall can move to help the water to run off, however due to the self-centering characteristic of the wall, it returns back to its original alignment after the event. As another instance, when a small relative movement of facing units is even critical, bonded PSRW can be considered to unite the facing units and limit the lateral movements.

5 Challenges and Future Research

Any new construction method has its own limitations and challenges, which need to be considered and studied for further development. Regarding the proposed pre-stressed segmental retaining wall system, experimental and analytical studies are required (for both unbonded and bonded cases) to investigate its potential and limitations. Practical aspects, such as the stage of construction and post-tensioning procedure need to be explored and its feasibility should be discussed by designers and construction companies. The geotechnical failure modes, soil-structure interaction and design criteria should be tested and validated. Design parameters such as pre-stressing loads and distance between the pre-stressing steel need to be explored. The behavior of the proposed systems in seismic areas should be investigated. The detailing of the proposed system must be addressed.

6 Conclusions

A new system of retaining walls named "pre-stressed segmental retaining walls (PSRWs), was proposed. The PSRWs system potentially has a collection of advantages over previously developed systems for construction of retaining walls. In particular, precast concrete/masonry segments can be incorporated which reduces the construction time with respect to conventional cantilever retaining walls. The method can also be incorporated in traditional MSWs to prevent local failure modes in those systems.

References

Ahmadi, H., Bezuijen, A.: Full-scale mechanically stabilized earth (MSE) walls under strip footing load. Geotext. Geomembr. **29**(2), 116–129 (2018)

Bathurst, R.J., Simac, M.R.: Geosynthetic reinforced segmental retaining wall structures in North America. In: Proceedings of the Fifth International Conference on Geotextiles, Geomembranes and Related Products, pp. 1–41 (1994)

Bean Popehn, J.R.: Mechanics and behavior of slender, post-tensioned masonry walls to transverse loading. Ph.D. thesis, The University of Minnesota, Minnesota, MN, USA (2007)

Bean Popehn, J.R., Schultz, A.E., Drake, C.R.: Behavior of slender, posttensioned masonry walls under transverse loading. J. Struct. Eng. **133**(11), 1541–1550 (2007)

Chang, K., Loh, C., Chiu, H., Hwang, J., Cheng, C., Wang, J.: Seismic Behavior of Precast Segmental Bridge Columns and Design Methodology for Applications in Taiwan. Taiwan Area National Expressway Engineering Bureau, Taipei (2002). Taiwan in Chinese

Chou, C.C., Chen, Y.C.: Cyclic tests of post-tensioned precast CFT segmental bridge columns with unbonded strands. Earthq. Eng. Struct. Dynam. 35(2), 159–175 (2006)

Chou, C.C., Hsu, C.P.: Hysteretic model development and seismic response of unbonded post-tensioned precast CFT segmental bridge columns. Earthq. Eng. Struct. Dynam. 37(6), 919–934 (2008)

ElGawady, M., Sha'lan, A.: Seismic behavior of self-centering precast segmental bridge bents. J. Bridge Eng. 16(3), 328–339 (2011). https://doi.org/10.1061/(ASCE)BE.1943-5592.0000174

Ganz, H.R.: Post-tensioned masonry around the world. Concr. Int. Detroit 25(1), 65–70 (2003)

Hassanli, R., Youssf, O., Mills, J., Fakharifar, M.: Analytical study of force-displacement behavior and ductility of self-centering segmental concrete columns. Int. J. Concr. Struct. Mater. (2017). https://doi.org/10.1007/s40069-017-0209-4

Hewes, J. T., Priestley N.: Seismic design and performance of precast concrete segmental bridge columns. Report No. SSRP-2001/25, Univ. of California at San Diego (2002)

Kerisel, J.: History of retaining wall design. In: Clayton, C.R.I. (ed.) Retaining Structures, pp. 1–16. Thomas Telford, London (1993)

Kim, T.H., Lee, H.M., Kim, Y.J., Shin, H.M.: Performance assessment of precast concrete segmental bridge columns with a shear resistant connecting structure. Eng. Struct. 32(5), 1292–1303 (2010). https://doi.org/10.1016/j.engstruct.2010.01.007

Koh, T., Hwang, S., Jung, H., Jung, H.: Settlement reduction effect of advanced back-to-back reinforced retaining wall. Int. J. Railw. 6(3), 107–111 (2013)

Kwan, W.P., Billington, S.L.: Unbonded posttensioned concrete bridge piers. I: monotonic and cyclic analyses. J. Bridge Eng. ASCE 8(2), 92–101 (2003)

Marriott, D., Pampanin, S., Palermo, A.: Quasi-static and pseudo-dynamic testing of unbonded post-tensioned rocking bridge piers with external replaceable dissipaters. Earthq. Eng. Struct. Dynam. 38(3), 331–354 (2009)

Ou, Y.-C., Chiewanichakorn, M., Aref, A.J., Lee, G.C.: Seismic performance of segmental precast unbonded posttensioned concrete bridge columns. J. Struct. Eng. 133(11), 1636–1647 (2007)

Tarawneh, B., AL Bodour, W., Masada, T.: Inspection and risk assessment of mechanically stabilized earth walls supporting bridge abutments. J. Perform. Constr. Facil. 32(1), 04017131 (2018)

Wight, G.D.: Seismic performance of a post-tensioned concrete masonry wall system. Department of Civil and Environmental Engineering, University of Auckland, Auckland, New Zealand (2006)

Waterproofing a Heterogeneous Soil (Sand-Bentonite) with Water and Leachate

Debieche Messaouda[1(✉)] and Mokadem Hassiba[2]

[1] LEGHYD F. de Génie Civil, Université des Sciences et de la Technologie Houari Boumediene, BP 32, El Alia/Bab Ezzouar, 16111 Alger, Algeria
mdebieche2015@gmail.com
[2] Faculté de Génie Mécanique et Génie des Procédés, Université des Sciences et de la Technologie Houari Boumediene, BP 32 El Alia/Bab Ezzouar, 16111 Alger, Algeria
hmokadz@yahoo.fr

Abstract. The existence of stocks of industrial waste, and household garbage on the surface poses major environmental problems. In Algeria, this problem begins to be felt, and becomes a concern of the environmental and geotechnical specialists, who care for the achievement of the sustainable developments. The primary interest for environment protection increases the requirement of high quality and reliable sealing systems, which the qui main characteristics Sought are a maximum dry density of the compacted sand mixtures/bentonite and a low hydraulic conductivity (K $\leq$ E-9 m/s) with water first and then with heavy metals (cadmium) to minimize leakage using a small percentage of bentonite.

1 Introduction

The interest in environment has recently considerably grown, and its protection is now included in the continuous action of the governments and the industries. In Algeria, the different types of waste are currently produced in almost 3.000 illegal dumps, occupying approximately 150.000 hectares. To overcome problem, waterproofing sites proves the most appropriate solution. It is a geotechnical method, to guard against the action of water or leachate, by narrowing the flow through a surface in a given time. The site's waterproofing technique, in the landfills sites, is nowadays a very necessary condition to protect the environment, which requires the use of appropriate materials.

The primary interest for environment protection increases the requirement of high quality and reliable sealing systems, whose the main characteristics sought are:

1. A maximum dry density of the compacted mixtures sand/bentonite.
2. A low hydraulic conductivity (K $\leq$ 10-9 m/s), to minimize leakage using a small percentage of bentonite.
3. Evolution of the concentration of cadmium through the mixture as a function of time.

In Algeria, the most important bentonite deposits are found in the west of Algeria in Fig. 1. They have been exploited since the 1950s.

M. Meguid et al. (Eds.): GeoMEast 2018, SUCI, pp. 197–208, 2019.
https://doi.org/10.1007/978-3-030-01944-0_15

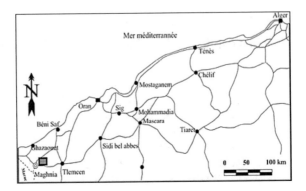

Fig. 1. Location map of the bentonite deposit in Algeria

In this context, first, an experimental campaign was conducted in order to define, the concentration of bentonite necessary to achieve a low permeability (K $\leq$ 10-9 m/s). The compaction characteristics and the possible variation of "K", allowing sustainability study of the required mixture. Second, a study of the evolution of the concentration of cadmium through the mixture as a function of time.

1.1   Materials Used

The type of bentonite used in this study is from Maghnia (West Algeria). It is marketed by the National Company of Mineral Products Non Ferrous Bental (ENOF), to be used in foundry, ceramic industry or the oil and hydraulic drilling.

1.1.1   Mineralogical Analysis
The X-ray diffraction pattern of the used bentonite, shown is in Fig. 2, show the montmorillonite is predominant compared to impurities such as quartz, orthoclase and some traces of calcite.

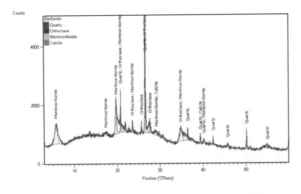

Fig. 2. X-ray diffraction pattern of the bentonite of Maghnia

Scanning Electron Microscope Analysis (SEM)

The analysis of the photography is taken by the SEM type JEOL JSM 6830.

On Fig. 3, it is shown that the bentonite is consisting of aggregates which are formed by assembling the montmorillonite particles of various sizes. This assembly is accompanied by the appearance of microcavities of different sizes.

Fig. 3. Photography of the Microstructure Bentonite of Maghnia

1.1.2 Physical Characteristics

The physical characteristics are determined from standard laboratory tests, according to the standards, as shown in Tables 1 and 2. The characterization tests show that bentonite is active montmorillonite, having consequently a relatively large water adsorption potential. This gives it a possibility of the use in reducing the permeability. Bentonite of Maghnia is sodic (noted SB). The used sand identification revealed that it is clean after the prewashing to remove all impurities, and bad graduated (noted Sp) in Fig. 4.

1.1.3 Permeability Tests

In general, determination of the coefficients of permeability to water, in the soil and-more generally, in the porous media- is based on the measurement of a flow rate, percolating through the microscopic porous structure, under the action of a positive or negative hydraulic load gradient. Free water and a capillary transfer governed by a mechanism, well reproduced by Darcy's law (Darcy et al. 1858)

Given that we are interested to the sealing of compacted mixtures, all samples made for permeability tests, will be compacted on the wet side, thus the permeability tests are carried out at water contents equal to: $\omega = \omega opt + 2\%$. SETRA/LCPC (1992).

Table 1. Results of identification tests of the studied bentonite

Identification ⟍ Designation	Bentonite of Maghnia
Natural water content (%)[1]	8,5
Specific density of solid grains. Gs [2]	2.72
Percentage of particles< à 80µm [3]	100
Percentage of particles C2< à 2µm [3]	42.5
Liquidity limit. Wl(%) [4]	240
Plasticity limit Wp(%) [4]	43
Plasticity index Ip (%) [4]	197
Consistency index Ic (%) [4]	1.18
Withdrawal limit Lr (%) [5]	10.24
Withdrawal index Ir (%) [5]	229.76
Activity (%) Ac [6]	6.06
Free swelling (ml) [7]	35
Value of methylene blue test [8]	18.75
specific area(m2/g) [9]	394
pH[10]	10.3

Table 2. Results of identification tests of the studied sand

Identification ⟍ Designation	Sand of Baghlia
Percentage of particles< à 80µm [3]	0.25
Specific density of solid grains Gs [2]	2.65
Effective diameter (mm) D10 [3]	0.26
Uniformity coefficient Cu [3]	3.46
Curvature coefficient Cc [3]	0.98
Equivalent of sand ES (%) [11]	97
Value of methylene blue test [8]	0.012

1: Determined according to the standard NF P 94-050 procedure
2: Determined according to the standard NF P 94-054 procedure
3: Determined as described in standard NF P 94-056 and NF P 94-057
4: Determined according to the standard NF P 94-051procedure
5: Determined according to the standard procedure ASTM D 427-61 and (XP P94-060-1)
6: Determined according to the standard NF P 94-051 procedure Ac = Ip/C2 or Ip/C2-n where Ac = (n = 5 if the soil is intact n = 10 if the soil is edited)
7: Determined according to the standard procedure ASTMD 5890 and planned in France NF P84 -703
8: Measure with methylene blue test (test spot) after NF P 94-068
9: Determine according to the standard procedure ASTM C 204-89
10: Determined a suspension of 20 g bentonite in 400 ml of distilled water
11: Determined according to the standard procedure NF P 18-0598.

The direct measurement of soil permeability according to (NF P94-512-11), is based on two so-called "constant load" for soils with high permeability or "variable load" for soils with low permeability. In this study, the two procedures are performed, since the mixture consists on two types of soil, namely sand and a family of clays

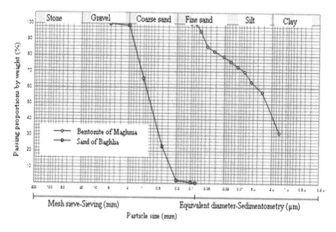

Fig. 4. Grain-size distribution curves of two materials used in this study

(montmorillonite). Once the optimum is obtained, from of the Proctor curve, the dry volumic weight and the corresponding content of water to be added are deduced. Another compaction test is carried out, this time, in the mold of Proctor permeameter, with a water content of the optimum +2% Fig. 5. The trials are conducted under the following conditions:

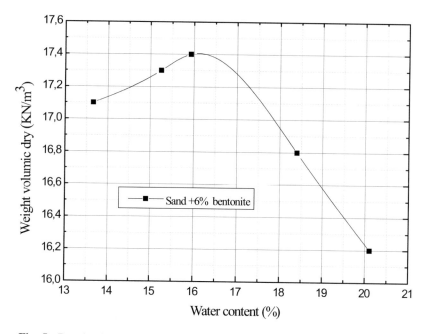

Fig. 5. Dry density versus the water content, for 6% concentrations of bentonite

1. The maintain of a low hydraulic gradient (noted i), during the saturation phase (i = 1).
2. The validity of Darcy's law;
3. The measurement of permeability after saturation, especially after allowing the bentonite to swell completely;
4. The temperature room is regulated at 20 °C, otherwise a corrections should to be made, according to the formula (2).

The device consists of a permeameter, for conducting the tests at constant and variable load simultaneously is presented on Fig. 6.

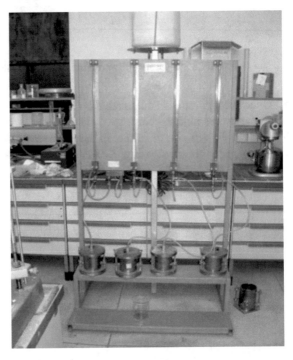

Fig. 6. Photography of the device of the permeability test with a permeameter of compaction.

It includes:

a. Four permeability cylindrical cell of 102 mm diameter and 117 mm height.
b. A panel of four glass tubes of 17 mm in diameter, which is connected to the four-cell permeability, for the variable load test.
c. A water tank for holding the specimen in the open air cell.
d. A tank used to fill the four tubes for precise level (h1).
e. Four metal rulers glued to the panel adjacent to the tubes, used to note the different water levels (h2).

Variable Load Test

For the variable load test, the permeability is given by:

$$K\left(cm/_s\right) = 2,3\,\frac{a.L}{A.T}.Log\frac{h_1}{h_2} \tag{1}$$

Where:

K: The permeability (cm/s)

a: The piezometric tube area (cm^2)

A: The sample area (cm^2)

L: The mold length (cm)

T: The time to go from h1 to h2 (s)

h1, h2: The successive levels water in the tube during a "T" time.

The permeability coefficient depends on the fluid viscosity: when the viscosity increases, the flow decreases, and vice versa. It is therefore appropriate to bring back the calculated values of permeability K, at a reference temperature, when the measurements are performed at various temperatures [16].

Conventionally, the coefficient of permeability "K" is given, for a temperature of 20 °C, where the water flows through the material, and the dynamic viscosity of water is of 1.00 MPa (AFNOR X30-441.2003).

$$K_{20\,°C} = K_{T\,°C}\,\frac{\eta_{T\,°C}}{\eta_{20\,°C}} \tag{2}$$

1.2   Déroulement de l'Essai de la Perméabilité Charge Variable

La procédure expérimentale se compose des essais de permeamètre à charge variable. La saturation et le gonflement sont effectués à l'intérieur du moule Proctor où le compactage du mélange sable-bentonite a été réalisé. Ces essais consistent à mesurer la variation de la perméabilité et la rétention des métaux lourds à savoir le cadmium à travers le mélange de sable-6%bentonite.

Le but de ces expériences est d'une part mesurer la perméabilité à l'eau, d'autre part, poursuivre, au bout du 15eme jour, l'essai de perméabilité sur les mêmes moules en passant à une eau polluée avec le cadmium à une concentration initiale de 50 mg/l afin d'évaluer le coefficient de perméabilité K.

- Mode opératoire

L'essai de perméabilité a été réalisé sur deux séries d'échantillons. L'une a contenu un mélange compacté à l'optimum +2% de l'optimum pour 6% de bentonite

1.3   Effects of Water on Permeability of Sand-6% Bentonite Mixture

After total saturation and stability of the swelling of mixtures made of 1 week duration at hydraulic gradient unit, we have proceeded with the assessment of the permeability coefficient to water in order to reach permeability according to the Algerian norm

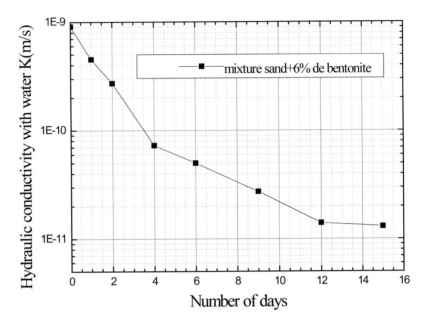

Fig. 7. Hydraulic conductivity versus time (water)

($K \leq$ E-9 m/s). Indeed, the curve in Fig. 7 illustrates the evolution of the hydraulic conductivity of the mixture over time. This evolution is progressive almost linear. The initial water permeability is 7.26 * E-10 m/s for the mix sand +6% of bentonite. It decreases by an order of magnitude to reach 3.24 * E-11 m/s and to stabilize after about twenty days from the first day of saturation.

1.4   Effects of Cadmium on the Permeability of Sand-6% Bentonite Mixture

The Fig. 8 illustrates clearly the evolution of the permeability coefficient after passing the cadmium in the samples as a function of time. Actually, for the permeabilities of sand-bentonite mixtures contaminated with cadmium (6% of bentonite), the increase was nearly 1 order of magnitude compared to the values obtained by the water permeability tests of the same mixtures. To give respectively a permeability of 5.53 * E-10 m/s and stabilizes by giving a final value of the hydraulic conductivity of 2.31 * E-10 m/s after about thirty days.

This increase is explained by the decrease of the swelling of the bentonite in the sand-bentonite mixture in function of the increase of the metal ions valence, which corresponds to a decrease in the thickness of the diffuse layer. The swelling of the bentonite is the element that gives the mixture its waterproofing properties according to the literature.

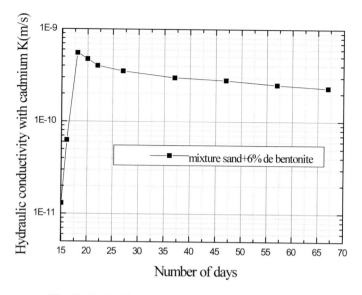

Fig. 8. Hydraulic conductivity versus time (cadmium)

1.5   Evolution de la Concentration en Fonction du Temps

Analyzes made at the Atomic absorption spectroscopy (AAS) after recovery of cadmium samples, show very low values considering the input concentration ($C0 = 50$ mg/l). This allows stating that we have an almost total elimination of pollutants by adsorption during a fortnight of the test passing metals through the moulds (Fig. 9). We can even argue that the saturation of the mould requires a lot of time.

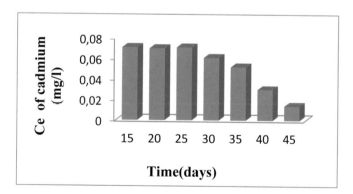

Fig. 9. Evolution of the cadmium concentration as a function of time (AAS method)

The first values are more important, this is explained by the fact that the mixture is preliminarily hydrated and saturated with water. The fact that the final concentration is low at the output shows the role of the bentonite resulted in the retention. That's to say that the bentonite has withheld the cadmium.

The results of the Atomic absorption spectroscopy (AAS) are confirmed by the scanning electron microscope and the diffractometer made on thin slides formed by the sand-6%bentonite mixture compacted at the end of the permeability test Figs. 10 and 11.

Fig. 10. Photograph illustrates the presence of cadmium in the mixture by MEB

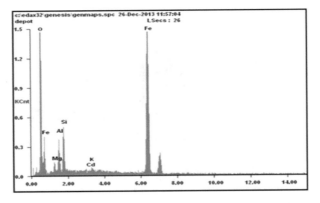

Fig. 11. Diffraction diagram of Sand-6% Bentonite mixture in the presence of Cd.

2 Conclusions

This study has two aims, at first to define the optimal composition of the sand-bentonite mixture using water that after compaction will have a permeability coefficient of less than or equal to 10-9 m/s and continue the permeability tests with the pollutant (Cd). It should be noted that the hydraulic conductivity after stability remains constantly less than or equal to 10-9 m/s.

On another hand, to try to work in the geo-environmental in order to analyze the physicochemical parameters such as the final concentrations that may, at a moment, weaken the sealing barriers (the sustainability of the structure). This has been con-firmed by the atomic adsorption spectrometer (AAS) tests which have shown that the Cd concentration becomes low over time, this concentration is acceptable according to Algerian regulations with permeability less than or equal to E-9 m/s and that the mould can perform for a long time. Therefore, the depollution is more effective with our adsorbent (Kettab 1992 and Boeglin 2000).

References

Académie des sciences: Contamination des sols par les éléments en trace les risques et leur gestion. Rapport 42 (1998)

AND: Décharges sauvages. Inventaires interprétation et recommandation, Octobre (2006)

AND: Revue de Presse n° 6 Octobre (2007)

BENTAL-ENOF: Etude technique et économique des gisements de M'Zila (Mostaganem) et Hammam Boughrara (Maghnia), vol I. Rapport Final. Bumigene.Inc, Montréal (2010)

Boeglin, J.-C.: Traitement et destinations finales des boues résiduaires. Techniques de l'ingénieur (2000)

Chalermyanont, T., Arrykul, S.: Compacted sand-bentonite mixtures for hydraulic containment liners. Songklanakarin J. Sci. Technol. 27(2), 313–323 (2005)

Commission de Normalisation AFNOR X 30P: Déchets-Perméabilité AFNOR/X 30P N 35 rev 1. X30-441: Détermination au laboratoire du coefficient de perméabilité à la saturation d'un matériau, document révisé lors de la réunion du 15/04/2003, 41 p (2003)

Costet, J., Sanglerat, G.: Cours des mécaniques des sols. Tome 1, 3ème édition, Ed Dunod, 285 p (1981)

D'Appolonia, D.J.: Soil-bentonite slurry trench cutoffs. J. Geotech. Eng. Div. ASCE 106(4), 399–418 (1980)

Darcy, H.: les fondations Publiques de la ville de Dijon, pp. 590–594. Victor Valmont Ed, Paris (1858)

Debieche, M.: Imperméabilisation d'un matériau pulvérisant par ajout de bentonite. Thèse de magister de la Faculté de Génie Civil, Université des Sciences et de la Technologie Houari Boumediene, Alger, Algérie (2007)

Debieche, M., Kaoua, F.: Characterization and valorisation of two bentonites in the waterproofing systems. Sci. Res. Mater. Sci. Appl. 5, 347–362 (2014)

Kettab, A.: Traitement des eaux (les eaux potables). Offices des Publications Universitaires, Alger (1992)

MATE: Plan National d'Actions pour l'Environnement et le Développement Durable (PNAE-DD) (2002)

MATE: Manuel d'information sur la gestion et l'élimination des déchets solides urbains (2003)

MATE: Atelier international sur le nouveau mode de gestion des déchets municipaux. Le centre d'enfouissement technique (CET) (2004)

MATE: a: Mise en œuvre du Programme National de Gestion des Déchets Ménagers. PROGDEM (2005)

MATE: Programme National de Gestion des déchets ménagers (2006)

Philipponnat, G., Huber, B.: Fondations et ouvrages en terre. Edition Eyrolles, 548 p (1997)

Robert, N.C.: Enjeux environnementaux et industriels-Dynamiques des éléments traces dans l'écosystème sol. In: spéciation des métaux dans le sol, les cahiers du club Crin, Paris, pp. 15–37 (1999)

Robinet, J.C., Rhattas, M.: Détermination de la perméabilité non saturée des matériaux argileux à faible porosité. Can. Geotech. J. **32**, 1035–1043 (1995)

SETRA/LCPC: Réalisation des remblais et des couches de forme – Fascicule I: Principes généraux. Guide technique du Ministère de l'équipement, du logement et des transports, p. 98 (1992)

Geotechnical Evaluation and Sizing of the Reinforcement Layer on Moles in the North Triangle Clover Project, in Brasília/DF, Brazil

Rideci Farias[1](✉), Haroldo Paranhos[1], Elson Oliveira de Almeida[2],
Joyce Maria Lucas Silva[3], Hellen Evenyn Fonseca da Silva[4],
and Lucas Inácio da Silva[4]

[1] Reforsolo Engineering, Brasília, Brazil
rideci.reforsolo@gmail.com, reforsolo@gmail.com
[2] UniCEUB, Brasília, Brazil
elsul0@hotmail.com
[3] CONCREMAT, Brasilia, Brazil
Joyce.civil@gmail.com
[4] UCB, Brasília, Brazil
heyennunl@gmail.com, lucas.ll.inacio@gmail.com

Abstract. The increase of the fleet of vehicles makes the Engineering is dynamic in solutions for the traffic flow. In Brasília, Distrito Federal, Brazil, it is no different. The city was designed for 500 thousand inhabitants, and currently, in the metropolitan region, circulate near 3 million. The growth of the city to a northern region, led to the development of a road complex called "Trevo de Triagem Norte" (TTN), composed of vehicles, road access, creation of new tracks and duplication of the Bridge of Bragueto on the lake Paranoá, giving access to the BR-020, as a solution to constant congestion. In the geopedological context, in the region, there are stretches with hydromorphic soils and water level at some times of the year. In geotechnical terms, soils are usually soft. Thus, this work presents a geotechnical evaluation of a region, but also the dimensioning of the reinforcement layer on the soils constitutes an end of future pathologies.

1 Introduction

This paper presents the results and analyzes of geotechnical studies carried out in the access road, under construction, at the Trevo de Triagem Norte (TTN) Project in Brasília/DF, with a view to the determination of soil resistance by means of standard tests Penetration Test (SPT) according to ABNT NBR 6484: 2001, Vane Test, according to ABNT NBR 10905 (MB-3122, October 1989), and the Light Dynamic Penetrometer (DPL) according to the recommendations of DIN 4094 and ISSMFE, 1989.

The studies consisted in the execution of a program of "Geotechnical Field Investigations" aiming at the evaluation of the soil resistance, considering that

© Springer Nature Switzerland AG 2019
M. Meguid et al. (Eds.): GeoMEast 2018, SUCI, pp. 209–220, 2019.
https://doi.org/10.1007/978-3-030-01944-0_16

geometric factors such as existing lateral ways and the lack of quota for the launching of deep drainage waters inhibited the traditional techniques of total soft soil removal, thus avoiding the use of temporary lowering and shoring of the excavations. Thus, a structural layer was designed to reinforce the subgrade, by means of the needling of the soft soil, without the need for longitudinal drains.

2   Location of the Area of Study and Performance of Tests

Figures 1, 2 and 3, taken from Google Earth on July 30, 2017, show approximate views of the study area.

Fig. 1. Macro view of the study area (Brasília/DF - work of the TTN).

Fig. 2. Macro view of the study area (Brasília/DF - work of the TTN).

Figure 4 shows the area in which the SPT, Vane Test, and DPL probes were tested for soil resistance. Area comprised between 2LW and 1WL.

The studies were carried out in a region with presence of soft soils at the site of implantation (Table 1). In the places where the work was already in place, the objective

Fig. 3. Micro view of the study area (Brasília/DF - work of the TTN).

Fig. 4. Area where the SPT, Vane Test and DPL probes were tested for soil resistance.

Table 1. Sites of study of soft soils.

Name		Site of soft soil		Construction phase
Site	Point	Stake	Side	
2LW	1	0 + 440.00	Right maple	In deployment
	2	0 + 520.00	Right maple	In deployment
	4	0 + 720.00	Right maple	In deployment
1WL	3	0 + 680.00	Left maple	In deployment
	5	0 + 380.00	Left maple	In deployment
	6	0 + 220.00	Right maple	In deployment
	7	1 + 0.00	Right maple	In deployment

was to evaluate the natural terrain close to the ground, in order to identify remnants of typical horizons of low supporting soil, adjacent to the implanted landfills.

In the places with difficulties of equipment penetration, due to presence of hardened surface layer, existence of compacted embankment, presence of interferences, among others, the reallocations of the holes between a radius of 2.0 m of the georeferenced points were made. Fieldwork was carried out in January 2017.

3 Scope of Services

Initially, the works were conducted in the office, with the evaluation of the results of the SPT surveys carried out in the area in question, in December 2016. Subsidized in these surveys, the incursion to the field was made on January 23, of the area and later for the Vane Test, DPL, and Trado surveys, and the collection of deformed and undisturbed samples, using Shelby, for the evaluation of the earthy mass. Figures 4, 5 and 6 show some of the activities developed in the field.

Fig. 5. Execution of probing.

Fig. 6. Execution of DPL probing.

Next, brief considerations are made on soft soils (Item 4), regarding landfills (Item 5), but also on consistency of the clays (Item 6).

After that, we present the results evaluation with the resistance profiles of the points studied by: "Vane-Test", Trado, SPT and the results obtained by the DPL.

4 Brief Considerations on Soils

In general, they are considered Moles, the high plasticity materials, in some cases rich in organic matter, with Nspt < 4. These are the deposits of organic soils, peat, very soft sands or hydromorphic soils, generally occurring in local as follows:

(a) low wetland areas;
(b) mangroves and swamps;
(c) Low-gradient river floodplains;
(d) old beds of water courses;
(e) Plains of marine or lake sedimentation.

According to the Brazilian Company of Urban Trains (CBTU), soft soils are those composed of very compressible materials unsuitable for support of landfill foundations, except in cases where special solutions such as balancing berms, vertical drains, slow construction of the landfill, etc. Among the soft soils, the following is highlighted:

(a) Clays almost always, but not necessarily organic, with soft to very soft consistency (SPT $\leq$ 2), apparent cohesion generally less than 1.50 t/m^2, natural moisture characteristically higher than the liquidity limit, or very close to it, a very high plasticity index (usually greater than 25%) and low permeability (k = 10–6 ak = 10–8 cm/s). In general, these clays are saturated or close to the saturation humidity and occur, almost always, flooded or with the level of the water table near the surface.

(b) Turfs of various types, commonly constructed by organic matter in a non-colloidal state, presenting a greater or lesser amount of remnants of partially or completely charred plant remains. In general, the peat is very permeable (k > 10–4 cm/s), low density, and also very compressible. A very important aspect to be considered is that the peatlands may not present themselves soft, in their present state of humidity, acquiring this characteristic when being wet.

5 Brief Landing Considerations (DNER-PRO 381/98)

In general, landfills are classified into three classes (Class I, II and III), according to the following characteristics:

• Class I: This class includes landfills along rigid structures, such as bridges and viaducts and other intersections, as well as landfills close to sensitive structures such as pipelines. The extent of Class I landfill shall be at least 50 m on each side of the intersection;

- Class II: landfills that are not close to sensitive structures, but are high, being defined as high ones that have heights greater than 3.0 m;
- Class III: Class III landfills are low, i.e. with heights less than 3.0 m away from sensitive structures.

6 State of Clays - Consistency

When handling a clay, one perceives a certain consistency, unlike the sands that easily collapse. For this reason, the state in which a clay is found is usually indicated by the resistance it presents.

The consistency of the clays can be quantified by means of a simple compression test, which consists of the compression rupture of a generally cylindrical clay specimen. The load that leads the specimen to the rupture, divided by the area of this body is called the simple compression strength of the clay (the simple expression expressed that the specimen is not confined, a procedure widely used in Soil Mechanics). As a function of the simple compressive strength, the consistency of the clays is expressed by the terms presented in Table 2.

Table 2. Consistency of the clays as a function of the compressive strength.

Consistency	Resistance, em kPa
Very soft	<25
Soft	25–50
Average	50–100
Hard	100–200
Very hard	200–400
Tough	>400

7 Evaluation of Results of Exploited Surpluses

The soft soils were classified according to Table 3, which relates several consistencies with the penetrations of the SPT (Percussion Sound), the penetrations of the PDL (Light Dynamic Penetrometer) and the "Su" resistance of the Vane Test test.

Thus, for the purposes of the evaluations herein, soft soils, the soils with characteristics of soft soils and very soft soils, are concomitantly considered. Table 4 shows the characteristics defined for the soft soils in question. Table 5 shows the depth of the soft soils, according to the surveys carried out in the area. It should be noted that the points marked with (*), in question.

Table 5, do not apply, because in these points, the soil profiles do not have characteristics of soft soils. They are siltous soils that in the natural state (confined) have good resistance, but when excavated or manipulated, they totally lose this resistance, not demonstrating their characteristics of very low resistance in the traditional polls. That is, SPT poles are described as sandy-loamy, variegated, hard-hardened soils with

Table 3. Variation of soil resistance.

Vane test	DPL	SPT	SPT	Consistency
Resistance (kPa)	N10	N30	N30	
Décourt, 1989	DIN 4094*	Lambe/Whitman Terzaghi/Peck	Godoy, 1972	
<25	0–3	0–2	0–2	Very soft
25–50	3–6	2–4	3–5	Soft
50–100	6–12	4–8	6–10	Average
100–200	12–22	8–15	11–15	Hard
200–400	22–45	15–30	15–19	Very hard
>400	>45	>30	>20	Tough

Table 4. Characteristics of soft soils.

Classification	N_{30} (SPT)	N_{10} (PDL)	Vane test (kPa)
Soft soil	<4	<6.0	<50

Table 5. Depth of soft soil.

Points of interest	Water level	Depth of soft soil (m)			
		Vane test	DPL	SPT	
Point 01	1.2	3.2	3.4	5.0	SP 0+440
Point 02	1.3	3.7	4.1	6.0	SP 0+540
Point 03	2.1	4.0	4.4	6.0	SP 01 and 02
Point 04	1.9	2.8	3.8	4.8	SP 05 and 06
Point 05	4.2	*	*	*	SP 07 and 08
Point 06	*	*	*	*	SP 11 and 12
Point 07	3.1	3.8	4.7	4.5	SP 0+640

SPT > 30. Therefore, in the CBR tests, they present CBR < 2%. On the other hand, the paving manuals suggest the replacement of these soils.

8 Size of the Substrate Reinforcement Layer

8.1 General Considerations

According to DNER-PRO 381/98 for stretches of soft soils less than 3.0 m thick it is economically feasible and complete removal of the soft layer, eliminating the problem entirely. However, whenever the thickness exceeds 3.0 m, more economical alternatives to living with the soft soil should be studied.

Since the Vane Test is the most suitable for the analyzes in question, it is evident that the layers of soft soils have variable "average" thicknesses from one section to another, with a minimum of about 2.8 m and a maximum thickness of 4.0 m. Considering the visits to the TTN deployment area and soil analysis in the region where there is low support capacity and high humidity of these soils, as well as the need for agility of the earthmoving services, it is recommended the partial removal of these soils until the depth and filling of the cavas with compacted granular stone material (needling). You should always observe the following immediate and proposed objectives over time:

(a) To structure the layer of the subgrade to stabilize the overlying layers;
(b) Uniformize the resistance of the surface layer;
(c) Serve as a draining layer in periods of high precipitation;
(d) Reduce the capillary strength of silt-clayey soils that in some cases can reach a few meters in height, revealing several pathologies in the pavements;
(e) To avoid the loss of soil resistance over time, caused by the gain or return of the soil moisture by means of the "field moisture balance", since the field humidity is above the optimum compaction humidity even in months considered drought;
(f) Optimize the schedule of earthmoving services, since good technique indicates that these services are rationalized in the dry periods, which normally ends in September.
(g) Adoption of a technique already consecrated in similar works of the region and that in general has a definitive character in the face of bad weather. Avoid the services of provisional lowering of the sheet, for the execution of earthmoving services;
(h) Avoid the provisional shoring of the excavations.

8.2   Determination of the Non-drained Design Resistance (SU)

According to the results of the Vane Test, the determination of the non-drained resistance (SU) of the project was determined by statistical means, according to the DNIT.

The DNIT has used the following sampling plan for the statistical analysis of the results of the tests:

$$X_{m\hat{a}x} = X + \frac{1.29\sigma}{\sqrt{N}} + 0.68\sigma \quad X = \frac{\sum x}{N}$$

$$X_{min} = \bar{X} - \frac{1.29\sigma}{\sqrt{N}} - 0.69\sigma \quad \sigma = \sqrt{\frac{\Sigma(X - \bar{X})^2}{N - 1}}$$

Being:

N = Number of samples
X = Individual value
$\bar{X}$ = Arithmetic mean
σ = Standard Deviation
X_min = Probable minimum value statistically

X_max = Statistically likely maximum value
N > 9 (number of determinations made)

Table 6 presents the summary of the non-drained (Su) strength values of the evaluated soft soils, and Table 7 presents the statistical analysis for the determination of minimum, medium and maximum values according to DNIT.

Table 6. Summary of undrained resistance values.

Survey	No. sample	S_u (kPa)
Hole 01	1	44.96
	2	27.07
	3	27.28
	4	24.45
Hole 02	5	31.43
	6	20.74
	7	14.84
	8	32.30
	9	44.96
Hole 03	10	42.34
	11	38.42
	12	40.60
	13	22.70
	14	16.81
	15	27.72
Hole 04	16	40.16
	17	28.38
	18	37.76
Hole 05	19	44.75
	20	29.47
	21	29.68
	22	20.74

Table 7. Statistical analysis of non-drained resistance values (Su).

Statistical analysis	
Values	Su (kPa)
Arithmetic mean	31.25
Standard deviation	9.31
Likely minimum	22.36
Probable maximum	40.15

8.3   Sizing of Sublayer Reinforcement Layer

For the design of the pavement structure, the formula proposed by Giroud and Noyrai was used to determine the height of the unreinforced road.

$$h_0 = \frac{1.6193x \log(N) + 6.3964x \log(P) - 3.7892xr - 11.8887}{(Cu)^{0.63}}$$

Being:

h_0 = altrua of the non-reinforced access road (m);
P = single-axle load of double wheels (kN);
r = allowable sinking on the road surface (meters);
N = number of passes of the vehicle;
Cu = undrained resistance of the foundation soil (kPa).

Using the appropriate parameters to determine h_0, the value of 1.73 m was found. According to the paving project, the following pavement structure is shown (Fig. 7):

Then, for the structure of the pavement in question, we have the sum of the thickness of the granular layer (sub-base + base) equal to 30 cm. Thus, the reinforcement layer of the subgrade will be equal to:

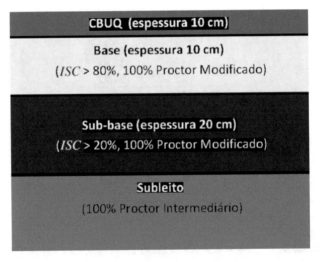

Fig. 7. Flexible floor structure.

hn = ho − (Base + sub-base thickness)

Being:
hn = Reinforcement thickness of the subgrade

So: hn = 1.73 m − (0.10 m − 0.20 m) = 1.43 m.

9   Conclusions

Subsidized in the information and results obtained, it is evident the presence of saturated soils and low support capacity (soft soils).

Among other techniques, the solution of partial removal of the soft soils, together with the needling with rock (launched stone rockfill) is a conventional practice that associates the desired stability to the executive schedules of this stage of the work.

In natural stretches the material is generally very soft to soft and in some cases with a layer with higher resistance. This resistance increase is coordinated by the dryness of the soil A horizon.

In general the solutions have to be feasible and bring the desired stability. Therefore, the reinforcement of the deeper layers and the control of capillary moisture must be considered in the proposed solutions.

In view of the results presented, it is evident the stabilization solution with a structuring layer with a crack, as it allows a lower ground needling by the passage of the earthmoving machines and with this the increase of the strength of the reinforced layer. The heterogeneity of the deformability of the lower layer is standardized by the "neg" presented before the working pressure used by the compaction equipment. Because of the variation of the layer resistance, the crack can penetrate more or less, according to Fig. 8. This type of service can be carried out both during rainy periods (with or without outcropping) and during periods of rainfall drought (Fig. 9).

Fig. 8. Sample withdrawal with Shelby.

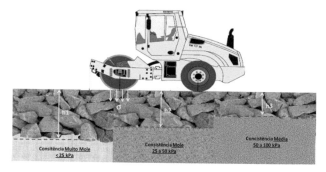

Fig. 9. Variation of penetration (needling) as a function of soil layer strength.

Acknowledgment

To the Consortium Via Conterc, Reforsolo Engenharia Ltda., Universidade Católica de Brasília (UCB), University Center of Brasilia (UniCEUB) and the Institute of Higher Education Planalto with contributions that made this work possible.

References

ABNT NBR 6484: Solo – Sondagem de simples reconhecimento com SPT – Método de ensaio (2001)

ABNT NBR 10905: Solo - Ensaios de palheta in situ - Método de ensaio (1989)

Companhia Brasileira de Trens Urbanos (CBTU) – IT – 116/CBTU: Execução de remoção e substituição de solo mole (2006)

Deutsches Institut für Normung. DIN 4094/International Society for Soil Mechanics and Foundation Engineering (ISSMFE): soil; exploration by penetration tests; aids to application, supplementary informations. Alemanha (1989)

DNER-PRO 381/98 – Projeto de aterros sobre solos moles para obras viárias

Evaluation of the Drainage Capacity of a Geocomposite Applied to the Green Roof of the Office of the Attorney General's Office in Brasilia, After 14 Years of Its Application

Haroldo Paranhos[1(✉)], Rideci Farias[1], Joyce Maria Lucas Silva[2],
Leonardo Ramalho Sales[3], Ranieri Araújo Farias Dias[4],
and Roberto Pimentel de Sousa Júnior[5]

[1] Reforsolo Engineering, Brasília, Brazil
reforsolo@gmail.com, rideci.reforsolo@gmail.com
[2] CONCREMAT, Brasília, Brazil
Joyce.civil@gmail.com
[3] UNICEUB, Brasília, Brazil
leonardoramalhosales@gmail.com
[4] Federal University of Pará, Belém, Brazil
ranierileislie@yahoo.com.br
[5] University of Brasília, UnB, Brasília, Brazil
eng.robertopimentel@gmail.com

Abstract. The purpose of this research was to verify such actions of the time in a geocomposite and to verify the general functioning of the system, thus allowing an in depth knowledge about the effectiveness of the use of this technological solution applied in drainage works. The research was carried out using a geocomposite from the suspended garden of 22,000 m^2 (green roof) of the building of the attorney general's office, located in Brasilia, after 14 years of its application. After the analysis, good stability of the drainage system was verified, through the evaluation of physical and biological clogging.

1 Introduction

In mid-2002, the Office of the Attorney General of the Republic was established for a 22,000 m^2 plant (Fig. 1).

The project, in fact, generated by the office of the famous architect Oscar Niemeyer, prior to the construction of a suspended garden (Fig. 2), thus the difficulty with the weight of the garden, the use of heavy equipment on the new waterproof phase and the illumination of the thickness of the soil layer necessary to the good vegetal performance; led to an engineering solution using geosynthetic materials (Fig. 3), replacing a 15 cm layer of gravel #02 with a 10 mm thick geocomposite.

This was done on request of 14 cm without vegetal soil, passing from 15 cm to 30 cm, improving the plant rooting, leaving some water in the drying fields.

M. Meguid et al. (Eds.): GeoMEast 2018, SUCI, pp. 221–231, 2019.
https://doi.org/10.1007/978-3-030-01944-0_17

Fig. 1. Garden on the slab in the Office of the Attorney General's Office in Brasilia

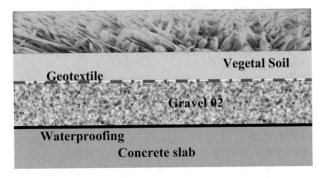

Fig. 2. Standard design.

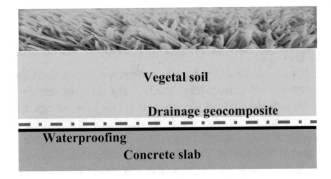

Fig. 3. Solution proposed with geosynthetics.

During 14 years the monitoring by the technical maintenance team was carried out in 2016 and it was submitted to a new technical evaluation in partnership with the body responsible for the execution and maintenance of the work in the epigraph.

They were evaluated as water infiltration abilities and response time, in the boxes and PVs.

2 Scope of Services

Initially the works were conducted in the office, with the evaluation of the existing projects of the work in question.

Subsequently, in possession of the projects, the system flow studies were carried out.

Based on the above mentioned studies, a field trip was done to carry out behavioral tests.

3 Location of the Study Area

Located in the South Federal Administration Sector (SAF Sul), near the Three Powers Square and the Esplanade of the Ministries, the headquarters of the Federal Supreme Court (STF), the Superior Court of Justice (STJ) and the Brazilian Court of Audit Union (TCU). Designed by Oscar Niemeyer in 1995, it was only ready in 2002, when the work was inaugurated.

Figures 4 and 5, taken from Google Earth on July 2, 2016, show approximate macro and micro views of the area of geotechnical studies and probed points.

Fig. 4. Macro view of the area of execution of geotechnical studies in Planaltina. DF.

Fig. 5. Micro view of the area.

4 Project Data

The project uses a drainage geocomposite applied on the slab of waterproofed concrete with asphalt and protected with a layer of 3.0 cm of mortar 1:3 (Cement:Sand).

The slab has a 1% drop in the form of "roof waters", where in the encounter (talvegue) the drainage drains of the drained waters are positioned (Fig. 6). Drainage water is these are collected and sent to the rainwater network by means of collector pipes positioned under the slab (in the garage).

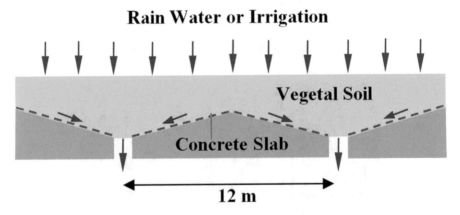

Fig. 6. Detail of the water drain of the drainage system.

5 Technical Evaluations Carried Out in the Area

5.1 Vegetation Assessment

The main indication that the drainage system is deficient is the health of the plants placed there. In general, the garden looks good, indicating an efficient drainage system (Fig. 7).

Fig. 7. View of the quality of the grass.

At some points, it is possible to notice a small difference in the staining of the grass, but this is an effect caused by the reflection of the sun's rays in the glass of the building, which causes this "burning" effect on the lawn.

According to the officials responsible for the maintenance of the garden, in all those years, since it was built, there was no significant problem with the drainage system, and it has been well maintained since its construction.

5.2   Evaluation of Infiltrations Under Concrete Slab

Evaluations were performed just below the slab, in the garage, no leakage points or infiltration of water from the drainage of the garden were detected (Fig. 8).

Fig. 8. View below the slab.

5.3   Assessment of Passage Boxes

All the in-place boxes were checked and opened for analysis (Fig. 9). In general, it was verified that 70% of the boxes were clean and free of organic soil or maize inside (Fig. 10) and 30% presented with soil and/or organic matter inside them (Fig. 11).

Fig. 9. Opening of the check boxes for visual verification.

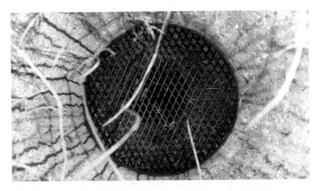

Fig. 10. Opening of the check boxes for visual verification.

5.4   Evaluation of the Drainage Nucleus

To evaluate the drainage nucleus a "window" was performed with the removal of the grass and the vegetal soil (Fig. 12). The geotextile was cut, exposing the draining nucleus (Figs. 13 and 14).

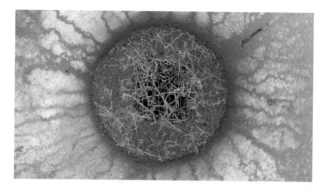

Fig. 11. Opening of the check boxes for visual verification.

Fig. 12. Opening windows on the lawn.

Fig. 13. View of the draining core free of soil.

Fig. 14. View of the drained core partially blocked.

5.5   Evaluation of Infiltration Capacity on Geotextile

Taking advantage of the cava (window - Fig. 12), the infiltration test was performed according to the ABNT NBR 1736 standard.

An average infiltration rate of 260 l/m^2 × day was found, which is equivalent to a permeability k = 3.01 × 10^{-4} cm/s.

5.6   Assessment of Grass Infiltration Capacity

For this evaluation, the flows of the implanted irrigation system were utilized, and the flow reached the wells of visits, related to that area of contribution (Fig. 15).

Fig. 15. Evaluation of the flow that arrives in the wells of visits.

In this evaluation, it was observed that the drainage system does not have an instantaneous outflow. In other words, for an irrigation volume of 6 mm /h, the system required almost 24 h for the total flow of the infiltrated water, without taking into account the quantity of water retained in the soil, nor evapotranspiration. In this context, a permeability of the system of $k = 1.67 \times 10^{-4}$ cm/s was found.

6 Draining Capacity Retroanalysis

Given the characteristics of the work we have:

$$\gamma = 16 \text{ kN/m}^3 \text{ (superficial soil)};$$

$$h = 0.30 \text{ m (thickness of soil on the geocomposite)};$$

$$i = 0.01 \text{ (Hydraulic gradient)};$$

For the calculation of the flow capacity of the Geocomposto we have. Vertical tension applied by the soil on the geocomposite:

$$\sigma = \gamma \times h$$

$$\sigma = 16 \times 0.30 = 4.8 \text{ kPa}$$

The Geocomposite flow rate for the working pressure is given by the interpolation of Table 1.

Table 1. Flow capacity of the geocomposite in the function of pressure.

Propriedade	Drenagem horizontal	Drenagem vertical	Método de Ensaio
Geocomposto FGC 1250/200			
Vazão [l/s.m]			
Pressão↓/Gradiente→ 0.01		1.0	ASTM D 4716
20 kPa	4.90×10^{-2}	7.93×10^{-1}	
50 kPa	4.69×10^{-2}	7.36×10^{-1}	
100 kPa	4.52×10^{-2}	6.97×10^{-1}	
200 kPa	4.29×10^{-2}	6.71×10^{-1}	

By interpolation of Table 1 we have:

$$Q_{\text{Geocoposto}} = 0.0501 \text{ l/s.m}$$

By the standard, the following reduction factors apply:

$$FR_{IN} = 1.20 \text{ (Soil intrusion)};$$

$$FR_{CR} = 1.20 \text{ (Flow - CREEP)};$$

$$FR_{CC} = 1.20 \text{ (chemical fouling)};$$

$$FR_{BC} = 1.15 \text{ (Biological filling)}.$$

However, since after 14 years, neither chemical nor biological fouling had been verified, they were adopted in the FRCC analysis = 1.00 and FRBC = 1.00.

Therefore, we have:

$$Q_{adm} = Q/(FRIN \times FRCR \times FRCC \times FRBC)$$

$$Q_{adm} = 0.0501/(1.20 \times 1.20 \times 1.00 \times 1.00)$$

$$Q_{adm} = 0.0358 \, l/s.m$$

Considering a rainfall of q = 0.03 (l/s)/m^2 (critical rainfall), we can determine the Flow that the drain will bear (Q_{Dren}):

$$Q_{Dren} = q \times E$$

$$E = \text{Drain length}.$$

Then:

$$Q_{Dren} = 0.03 \times 6$$

$$Q_{Dren} = 0.18 \, (l/s)/m$$

If we associate that the Security Factor can be expressed by FS = Q_{adm}/Q_{dren}; then we have:

$$FS = 0.0358/0.18$$

$$FS = 0.20$$

7 Analysis of Results

Although the geotextiles were apparently employed by soil layer, no signs of physical clogging were detected in the geotextile, to the point of preventing the flow of water.

There were no damages caused by possible rainwater or irrigation water soluble materials, which could have caused damage to the drains, thus damaging their ability to function.

Some visitor wells were partially clogged with soil. This fact was due to the surface runoff, increased by the thinning of the superficial soil layer, normally found in the soils of the region, used as surface soil erroneously called vegetal or organic soils. They are in fact Latosols that form a carapace in the face of cycles of wetting and drying, together with the presence of chemicals from irrigation (Fig. 16).

Fig. 16. Hardened surface layer.

The fact that a Safety Factor (FS = 0.20) was found, well below the usual drainage values (FS between 2 and 5), the system has been in operation for all these years. This bottleneck in the drainage system justifies the time needed to drain the area (24 h), observed in the Infiltration Capacity Evaluation on Grass Item 5.6.

8   Conclusions

After visiting and checking the general conditions it was possible to attest to the perfect functioning of the drainage system. This indicates that the choice of the use of geocomposite or the substitution to the conventional methods with gravel, not only generated decrease of the work costs, of the execution time, smaller but also the effectiveness of the use of geosynthetics in this type of system. Relating to the duration of materials they continue to function at their full capacity, even after several years following their application, requiring improvements through hardened surface layers.

It is observed that even presenting a below-normal safety factor, the system can infiltrate the rainwater, accumulate in the voids present in the soil layer and gradually drain through the proposed system, generating a delay in the final release of rainwater.

Abstract: The Long Term Performance of Geocomposite Drainage Materials Used as Capillary Break Layers

Alain Hérault[1]([✉]) and Dave Woods[2]

[1] Low & Bonar, Paris, France
alain.herault@lowandbonar.com
[2] Low & Bonar, London, UK
dave.woods@lowandbonar.com

Abstract. In arid regions, groundwater is often saline and the water table can be relatively close to the surface. Capillary rise above the water table brings dissolved salts near to the surface of the ground which can result in severe problems for both vegetation and the foundations of building structures.

For vegetation to survive, it is essential that there is a barrier between the saline ground and the clean ground and that this barrier also provides good drainage. A capillary break layer is installed above the highest level of the water table so that the void is never totally saturated by ground water and that the capillary break has high in-plane flow capacity to remove excess precipitation and irrigation water that enters through the clean ground above.

Chloride ions within saline moisture can be drawn into stone and concrete where the resultant chemical reaction causes expansion and weakening of foundations and the degradation of architectural finishes such as marble flooring. An effective barrier is to provide a capillary break layer between saline ground and clean ground. The capillary break layer creates a void across which capillary rise of saline moisture is prevented.

In both instances it has become common to use geosynthetic drainage layers to provide this capillary break. This paper introduces the application of capillary break layers and discusses in detail the need for ample understanding of the in situ and long term performance characteristics of the different types of drainage composite.

1 Capillary Break Layers

In coastal areas and particularly in arid regions like the GCC the ground water is often saline with a high concentration of chloride ions. In addition to the problems this poses for provision of drinking water the chloride ions can be harmful to vegetation and manmade structures. Salt concentrations of only 4 or 5 g/l are generally too high for most plant species to survive whilst the salts can damage and discolour decorative marble floorings and paved areas and cause concrete in foundations and basement structures to deteriorate.

The phenomenon of capillary rise can cause chloride ions to rise up 2 m above the high water table within the soil necessitating the use of precautionary measures to

© Springer Nature Switzerland AG 2019
M. Meguid et al. (Eds.): GeoMEast 2018, SUCI, pp. 232–242, 2019.
https://doi.org/10.1007/978-3-030-01944-0_18

protect masonry, concrete and vegetation which may be affected. The traditional methods included both the protection of vulnerable concrete with bituminous paint or chemical admixtures as well as the use of free draining granular layers, usually 200 mm thick or greater, with very low fines contents to act as a discontinuous void within the soil strata to prevent the capillary rise.

2 Geocomposite Drainage Layers

Geocomposites were first developed in the 1970's as the geosynthetic industry began to establish itself and broaden the offering from geotextiles. Drainage was one of the first applications for a geocomposite with the pioneering development beginning with a filament core structure. Currently there are many variations of drainage composites available with widely varying core types all using a geotextile filter on one or both sides. The polymer types used also vary with PP, PA or PET being most commonly used. PP and PA tend to be more stable at a molecular level giving better durability in most soil environments whereas PET is susceptible to high alkaline levels which results in deterioration of the properties through hydrolysis.

The form of the drainage composites typically consists of two layers of geotextile filter fabric on either side of an open core structure which usually consists of 3D filament structures, cuspated thermoformed sheets, box like structures or geonets formed of two layers of extruded polymer to form a grid.

All the various types of drainage geocomposite available function by maintaining an open void that allows drainage along the material whilst the filter layers prevent the material becoming clogged with the migration of fine soils. They are typically all designed with similar in plane flow capacities to provide similar drainage to a 500 mm stone drainage layer.

These materials are typically less than 2 cm thick, very light weight, easy to transport and install. An added benefit is that geosynthetic solutions will typically have 80% less carbon footprint than the traditional solution of a 20 cm gravel layer due to the reduced need for quarried fills, additional transportation, excavation and disposal of spoil from site. For salt barrier application the filament core lends itself well as the 95% open structure better ensures a void break to stop the saline flow while still providing adequate discharge from any excess irrigation or precipitation flow whilst the flexibility of the fabric allows it to be bent and formed on site within planting trenches or around foundations.

3 The Practical Application of Capillary Breaks

A capillary break should be situated above the HW level so it does not become saturated by the contaminated, saline water, and must have sufficient flow capacity that any irrigation or ground water flow percolating down to the salt barrier is drained away through the geosynthetic salt barrier without restriction.

A 6 mm thick, filament "V" shaped core structured geosynthetic salt barrier is particularly good at creating a void break as it is 95% air space and will allow water to flow in all directions, the filter used on both sides of the core may be adapted to suit the particular soil types encountered. This type of salt barrier while having a very open structure is also able to withstand the high load environments, such as below paved road structures, without damage or fatigue. This can be particularly important during installation when often site vehicles drive over the geosynthetic materials with insufficient fill material on top and can cause, in the case of cuspated core structures, permanent damage to the core thus reducing its flow capacity.

In 2014/15 a project of 800,000 m^2 of salt barrier was constructed in the GCC to prevent discoloration of the marble surface used for pedestrian pavements and vehicular access roads (Rawes and Holtus 2015). In this project Enkadrain 5006H/5-2 s/M200PP, a 6 mm geosynthetic "V" shaped filament core with 200 g nonwoven filters on either side was used to create the void break and provide filter/separation interfaces. Ideally the salt barrier should be laid at a gradient of 1–3% to ensure there are no back falls. In this project any excess irrigation or precipitation water runoff was drained away to subsurface drainage, therefore the drainage design capacity of the salt barrier was based on the saline capillary flow. These flows are clearly much lower than the precipitation drainage and rarely present a design issue. The flow capacity of the "V" shaped filament core was 1.7 l/s m at 20 kN/m^2 R/F loading.

In this project the material used was 5 m wide by 100 m long rolls which for such a large area made installation much faster. To ensure well covered joints the filter fabrics are 100 mm wider than the core material thus allowing an overlap of the filter fabric at each joint top and bottom. For joints in the length direction a 100 mm of the core structure is cut out between the filters such that the next roll may be inserted into the slot thus providing an overlap top and bottom (Figs. 1, 2, 3 and 4).

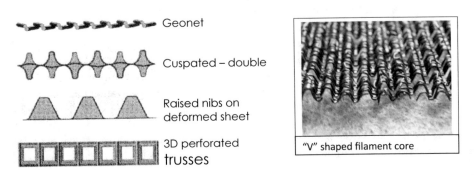

Fig. 1. Typical types of drainage geocompsite

Vehicles must not drive directly on any geosynthetic, usually before site vehicles are allowed to drive over the construction a 300 mm minimum (dependent on site compaction equipment used) thick fill layer on top of the salt barrier is required.

On the same project there were also areas for planting palm trees and shrubs which also needed protection from the damaging salts rising up from the HW level. In this case the trenches excavated for the palm trees or shrubs were lined on all sides and the

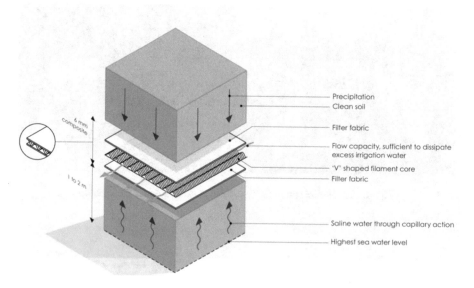

Fig. 2. The positioning of a capillary break layer

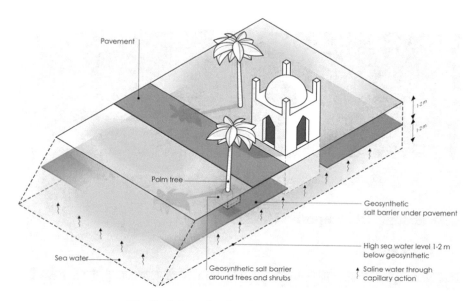

Fig. 3. Schematic of capillary break locations

bases with the "V" shaped filament core (ref Fig. 5 below) salt barrier ensuring interception of the saline salts before they reached the roots of the trees. Even saline levels of 4 or 5 g/l are generally too high for plants and trees to survive. The salt barrier also provided drainage to any excess irrigation water which would then be piped away from the salt barrier.

Fig. 4. Installation procedures on site and edge detail for overlap

Fig. 5. Detail of planting pit barriers around tree root balls

4   Drainage Geocomposite Design

4.1   Input Flow Rate

Usually, no specific input flow rate is given in this type of application, the design of the drainage composite is typically based on the performance of the more traditional drainage layer. The discharge capacity of the gravel layer is calculated with the help of Darcy's law, by considering flat slopes only, in line with the capillary break application. It depends on the maximum flow length, the hydraulic conductivity (k-value) and the thickness of the considered granular layer. Typically, drainage geocomposites are designed to have comparable drainage capacities to a granular layer 500 mm thick with a hydraulic conductivity of 10^{-3}–10^{-4} m/s which is far in excess of the values needed in the capillary break application. (Comité Français des Géosynthétiques 2014)

4.2   Specific Design Parameters for Geocomposites

The water flow capacity of a geocomposite depends on the degree and duration of compressive stress, the hydraulic gradient of the flow and the boundary conditions. In a capillary break application, the compressive stress is given by the dead load of the upper soil layers and, when relevant, the applied live loads. The hydraulic gradient is defined as horizontal as we are typically not considering any water pressure or head on top of the geocomposite although for the collection and dissipation of percolated rainwater and excess irrigation water there is typically a 1–3% fall in the installed fabric towards the site collector drain.

Since the geocomposite is installed between two soil layers, the discharge capacity of the geocomposite has to be measured in accordance with the ISO 12958 with the correct boundary conditions. Depending upon the application, discharge capacity can be measured on specimens installed between two foam layers (F/F option) in order to simulate the deformation of the surrounding soil layers, or on specimen installed between foam layer on one side and rigid plate on the other side (R/F option). The R/F option is only relevant for applications in contact with HDPE geomembranes or concrete surface on one side of the composite. Discharge capacity can also be measured on specimens installed between two rigid plates (R/R option) in specific case of leakage detection between two HDPE geomembranes or when the geocomposite is located between shotcrete and structural arch in tunnels. In the case of capillary breaker application, the boundary conditions in the in-plane water flow capacity test should normally be two foam layers (F/F option). It is important when selecting such products that comparisons of drainage capacity are made for similar conditions.

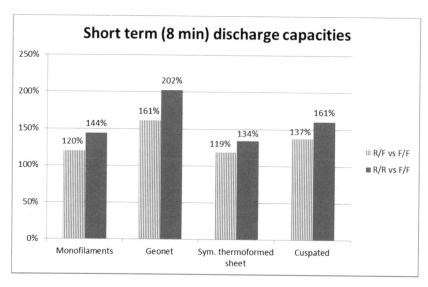

Fig. 6. Synthesis of overestimation of the in plane-flow capacity depending on the conditions of test (Touze Foltz et al. 2014)

The impact of this last parameter on the short term performances under 200 kPa of several types of geocomposite structures was described by Touze Foltz et al. 2014. Figure 6 shows, per product type, the test conditions effect on the discharge capacities obtained after 8 min for the four materials tested between two rigid plates (R/R option), between a rigid plate and a standardized foam layer (R/F option), versus the performances measured between two foam layers (F/F option). The flow rate value at 8 min corresponds to the time of the first measurement given in EN ISO 12958.

The research indicated that the short term discharge capacity could be overestimated by between 120% and 200% depending on the type of core structure when the wrong boundary conditions were selected for the testing.

4.3   Long Term Performance

The long term function of geocomposite drains is essential to ensure the salt barrier is not compromised during the service life, typically 50–100 years.

Flow reduction with time is considered through the following reduction factors:

– compressive creep(RF_{CR-Q}), i.e., time-dependent in-plane discharge capacity reduction due to thickness reduction of the draining core under the applied stress;
– delayed intrusion (RF_{IN}), i.e., decrease of the in-plane discharge capacity over time due to geotextile intrusion into the draining core resulting from time-dependent deformation of the geotextile filter (Fig. 7);

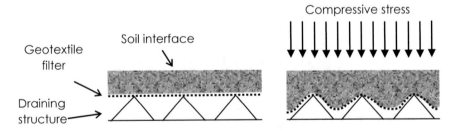

Fig. 7. Intrusion phenomenon

There are several processes to assess the reduction factors RF_{CR-Q} and RF_{IN} described above.

The first process, suggested by NF G38-061 (Afnor 2017), is to simultaneously assess (RF_{CR-Q} x RF_{IN}) these two parameters, by performing a long term laboratory test with the appropriate boundary conditions. This approach requires either a specific hydraulic device to measure over time the long term decrease of the water flow capacity of the tested specimen compressed at a specific normal load and for a specific hydraulic gradient, or, for very long term test (i.e., several years), compressive creep boxes applying appropriate boundary conditions in parallel with the usual water flow capacity test device in accordance with EN ISO 12958.

Low & Bonar have for many years measured these parameters in lab equipment created specifically to recreate these conditions (Böttcher 2006) to predict the long term performance of its drainage composites on the basis, for some products, of a test period of up to approx. 125 months (>10 years), by taking into account simultaneously the thickness reduction and long term intrusion phenomenon. The compressive creep tests are carried out in blue boxes as shown in Fig. 8a. They were designed to carry out creep tests with several types of contact surfaces (Rigid/Rigid, Rigid/Flexible or Flexible/Flexible) under normal loads up to 400 kPa.

(a)

(b)

Fig. 8. Photos showing the setup used in the creep tests

The specimen are loaded by means of pressure bags which are placed on top and bottom part of the box. A flexible support is realized by direct contact of the sample with a natural rubber membrane of 1 mm thickness, a rigid support is realized by metal plates kept between the samples and the natural rubber membrane. The air pressure on the top and bottom side of the box is monitored using manometers.

The water flow capacity tests are carried out in device (Fig. 8b) as required by the test standard EN ISO 12958 "Geotextiles and geotextile-related products – determination of water flow capacity in their plane". The water flow capacity of the specimen is tested under the following conditions:

– same stress as the one applied in the compressive creep test box,
– hydraulic gradient i = 1.0 or 0.1
– Rigid/Foam or Foam/Foam contact surfaces in accordance with EN ISO 12958,

The specimen is reinstalled in the compressive creep test box after each water flow capacity measurement. Figure 9 shows an example of long term water flow capacity curve including thickness reduction and continuous fleece intrusion on long term of semi-compressible product with V-shape monofilament draining structure.

The second process consists in assessing separately each one of the both reduction factors RF_{CR-Q} and RF_{IN}. Several methods are proposed worldwide.

The first method predicts the water flow capacity of a drainage geocomposite at the end of the service life by considering the assumption that the flow capacity is directly dependent on the residual thickness at the end of the service life as predicted by the extrapolation of compressive creep tests carried out between Rigid/Rigid support conditions (EN ISO 25619-1). This residual thickness is therefore used to determine, with

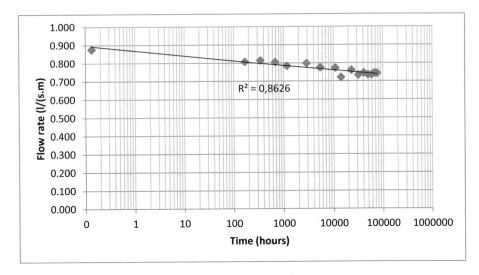

Fig. 9. Long term water flow capacity curve, including thickness reduction and continuous fleece intrusion phenomenon during 76 000 h in the blue boxes.

the help of the short term compressive curve of the geocomposite (EN ISO 9863-1), the normal stress which has to be applied in a short term water flow capacity test, carried out according to EN ISO 12958 with the requested boundary conditions. According to this method, this latest short term test simulates the long term water flow capacity of the composite at the end of the service life, considering intrusion phenomenon as a short term phenomenon taking place during the water flow capacity test (8 min) only. This assumption of intrusion being a short term response only does not consider any reduction factor for delayed intrusion (RF_{IN}) but introduces a specific reduction factor (RF_L) for overall uncertainties on laboratory data and field conditions.

Measurements of water flow capacity of drainage composites according to EN ISO 12958 with Foam/Foam option, after loading for 168 h under a normal stress of 100 kPa with, during compressive creep stage, either Rigid/Rigid support conditions in one case or Foam/Foam support conditions in the second case, have shown that the actual decrease in water flow capacity may be significantly higher in the second case, particularly for incompressible draining core structure like a geonet (Stoltz and Hérault 2016) where the geotextile filter fabrics have been witnessed to come into contact in the regular voids of the net under prolonged creep intrusion. The ratio $Q_{t=168h}/Q_{t=0}$ is equal to 1.0 in case of Rigid/Rigid support conditions for creep stage process, this confirms there was no thickness reduction of the draining core during 168 h in the case of a geonet, filter intrusion phenomenon being ignored in Rigid/Rigid support conditions. The ratio $Q_{t=168h}/Q_{t=0}$ falls down to 0.42 in case of Foam/Foam support conditions for creep stage process (Fig. 10). This example leads to a RF_{IN} value of 2.38 at t = 168 h under 100 kPa and RF_{CR-Q} = 1.0 (incompressible draining core) and shows that short term water flow capacity test with Foam/Foam option after 168 h of compressive creep

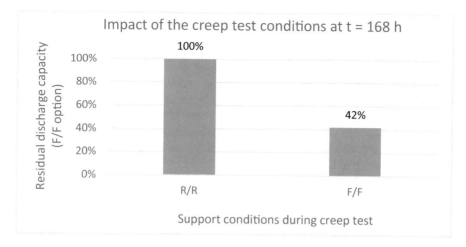

Fig. 10. Water flow capacity (Foam/Foam, i = 1.0, 100 kPa) decrease of rigid geonet for two compressive creep support conditions (Rigid/Rigid – Foam/Foam) at t = 168 h.

carried out between two rigid plates overestimate by a factor of 2.38 the actual water flow capacity of the geocomposite loaded through two foam layers during the creep stage.

To evaluate the suitability of the drainage composites for the conditions on a given site, attention should be given by the designers to the method used to assess the long term performance of geocomposite, particularly when they are used as capillary break layers or in any other application where the drainage composite is installed between two soil layers and in an application where any potential bridging of the void due to soft soil intrusion may allow capillary action through the material.

Additional attention should be paid to the failure mode of geocomposites when the long-term flow capacity is predicted. So-called incompressible geospacers can collapse at higher long term load-levels, particularly if they have been exposed to shear forces such as vehicles slewing their tracks above the textile. In order to ensure that a capillary break layer retains sufficient drainage capacity and does not suffer a collapse which might lead to a loss of its function the geocomposite must be tested in the appropriate long term conditions to ensure its service life.

Long term performances of the drainage composite up to the end of the service life should also be checked against the risk of biological and chemical clogging which could lead to a flow rate reduction over time as well although this is less relevant in capillary break applications where the material will be dry for the majority of the service life.

5 Conclusion

The increased usage of geocomposite salt barrier/capillary break layers demonstrates the growing confidence of the Engineering industry in the technology and allows them to realise the benefits in time and cost savings and reduction in the environmental impact of their projects. The variety of material types available and the confusion that can be caused when attempting to compare product technical data sheets can be off putting however, more work is needed to ensure common test procedures are used and reported values are consistent. Additionally on site monitoring of drainage performance in the long term would serve to demonstrate the viability and longevity of the techniques.

This paper hopefully serves to clarify the impact of the boundary conditions over the in-plane water flow capacity of the drainage geocomposites, particularly in their application as a capillary break layer where flow rate over-estimation due to testing in inappropriate boundary conditions may allow the drainage composite to be over-whelmed by precipitation or irrigation water but more specifically where soft soil intrusion or core collapse could allow the filter layers to contact each other thereby removing the capillary break function.

References

AFNOR 2017. NF G38-061 Use of geotextiles and geotextiles related products - drainage and filtration systems - justification of dimensioning and design elements

Böttcher, R.D.: Long-term flow capacity of geocomposites. In: Proceedings of 8th International Conference on Geosynthetics, Yokohama, Japan, pp. 423–426 (2006)

Comité Français des Géosynthétiques: Recommandations pour l'emploi des Géosynthétiques dans les systèmes de Drainage et de Filtration (2014)

Rawes, B., Holtus, R.: Salt Barrier Applications in the Middle East (2015)

Stoltz, G., Hérault, A.: Long term filter intrusion phenomenon in several types of drainage structures. In: Euro-geo 6, Ljubljana, Slovenia, pp. 575–583 (2016)

Touze Foltz, N., Hérault, A., Stoltz, G.: Evaluation of the decrease in long term water flow capacity of geo-composites due to filter intrusion. In: 7th International Congress on Environmental Geotechnics, Melbourne, pp. 321–329 (2014)

Carbon Footprint of Geomembrane Alvatech HDPE vs Traditional Waterproofing Barrier

Jose Miguel Muñoz Gomez[✉]

Geosynthetics Division, SOTRAFA SA, Almería, Spain
jmm-geo@armandoalvarez.com

Abstract. Lowering the Carbon Footprint is one of the strongest advantage by using HDPE geomembrane instead of traditional way for waterproofing as compacted clays.

A HDPE liner 1.5 mm could give similar watertight as 0.60 m compacted of high quality and homogeneous clay with lower permeability than $1 \times 10{-}11$ m/s (ASTM D 5887). Based on several scientist survey, considering all resources and energy to become either products as a waterproofing barrier, the geosynthetics (geomembrane HDPE 1.5 mm) takes up lower carbon dioxide equivalent, therefore it is more environmentally friendly solution.

1 Introduction

Features of HDPE Geomembrane and its Carbon Footprint.

The main component of HDPE is the monomer ethylene, which is polymerized to form polyethylene. The main catalysts are aluminum trialkylitatanium tetrachloride and chromium oxide.

The polymerization of ethylene and co-monomers into HDPE occurs in a reactor in the presence of hydrogen at a temperature of up to 110° Celsius degrees (230 degrees Fahrenheit). The resulting HDPE powder is then fed into a pelletizer to make pellets.

Then, SOTRAFA, as a manufacturer with latest technology in calandred system (flat die), makes geomembrane ALVATECH HDPE from these pellets. The Geomembrane ALVATECH HDPE keeps its outstanding features constantly either dry season or wet season.

1.1 GHG Identification and CO_2 Equivalents

The GHGs (Greenhouse Gas protocol) included in the calculation were the three (3) primary GHGs, namely carbon dioxide, methane, and nitrous oxide. Each of these gases has a different Global Warming Potential (GWP), which is a measure of how much a given mass of a greenhouse gas contributes to global warming or climate change.

Carbon dioxide is by definition issued a GWP of 1.0. To quantitatively include the contributions of methane and nitrous oxide to the overall impact, the mass of the methane and nitrous oxide emissions are multiplied by their respective GWP factors

© Springer Nature Switzerland AG 2019
M. Meguid et al. (Eds.): GeoMEast 2018, SUCI, pp. 243–247, 2019.
https://doi.org/10.1007/978-3-030-01944-0_19

and then added to the mass emissions of carbon dioxide to calculate a "carbon dioxide equivalent" mass emission. For purposes of this paper, the GWPs were taken from the values listed in the USEPA regulations "Mandatory Reporting of Greenhouse Gas Emissions" US EPA (2009). The GWPs for the GHGs considered in this analysis are:

- Carbon dioxide = 1.0 GWP 1 kg CO_2 eq/kg CO_2
- Methane = 21.0 GWP 21 kg CO_2 eq/kg CH_4
- Nitrous oxide = 310.0 GWP 310 kg CO_2 eq/kg N_2O

Using the relative GWPs of the GHGs, the mass of carbon dioxide equivalents (CO_2eq) was calculated as follows:

(1)   kg CO_2 + (21.0 × kg CH_4) + (310.0 × kg N_2O) = kg CO_2 eq (Figs. 1 and 2)

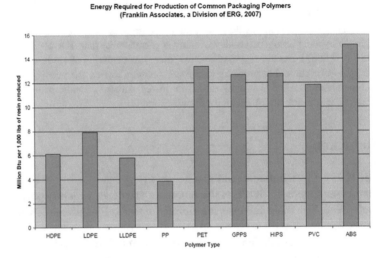

Energy Required for Production of Common Packaging Polymers
(Franklin Associates, a Division of ERG, 2007)

6 million Btu = 156 liters fuel equivalent = 1,758 KW*hour

Fig. 1. Energy required for production of common packaging polymers (Franklin Associates, a Division of ERG, 2007)

1.1.1 Assumption
The energy, water, and waste information from the extraction of the raw materials (oil or natural gas) through production of HDPE pellets and then manufacturing geomembrane HDPE:

- 1.5 mm thick HDPE geomembrane, with density 940 kg/m^3
- HDPE carbon footprint is 1.60 kg CO_2/kg polyethylene (ICE 2008).

(2)   940 kg/m^3 × 0.0015 m × 10,000 m^2/ha × 1.15   (scrap   and   overlaps) = 16,215 kg HDPE/ha

(3)   E = 16,215 kg HDPE/ha × 1.60 kg CO_2/kg HDPE => 25.944 kg CO_2 eq/ha

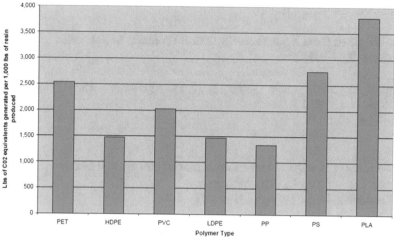

Fig. 2. Greenhouse gas emissions generated in polymer production (Franklin 2007; Vink et al. 2007)

1.1.2   Assumption Transport: 15.600 squm/Truck, 1000 km from Manufacturing Plant to Jobsite

- 10.15 kg CO_2/gal diesel $\times$ gal/3,785 l = 2.68 kg CO_2/l diesel
- 0.26 g N_2O/gal diesel $\times$ gal/3,785 l $\times$ 0.31 kg CO_2 eq/g N_2O = 0.021 kg CO_2 eq/l diesel
- 1.44 g CH_4/gal diesel $\times$ gal/3,785 l $\times$ 0.021 kg CO_2 eq/g CH_4 = 0.008 kg CO_2 eq/l diesel

On-Road truck product transport emissions:

(4) E = TMT $\times$ (EF CO_2 + 0.021·EF CH_4 + 0.310·EF N_2O)

(5) E = TMT $\times$ (0.972 + (0.021 $\times$ 0.0035) + (0.310 $\times$ 0.0027)) = TM $\times$ 0.298 kg CO_2 eq/ton-mile

Where
E = Total CO_2 equivalent emissions (kg)
TMT = Ton Miles Traveled
EF CO_2 = CO_2 emission factor (0.297 kg CO_2/ton-mile)
EF CH_4 = CH_4 emission factor (0.0035 g CH_4/ton-mile)
EF N_2O = N_2O emission factor (0.0027 g N_2O/ton-mile)
Converting to Metric Units

(6) 0.298 kg CO_2/ton-mile $\times$ 1.102 tons/tonne $\times$ mile/1.61 km = 0.204 kg CO_2/tonne-km

(7) E = TKT $\times$ 0.204 kg CO_2 eq/tonne-km

Where
E = Total CO_2 equivalent emissions (kg)
TKT = tonne – km Traveled

1.1.3 Distance from Manufacturing Plant (Sotrafa) to Job Site (Hypothetical) = 1000 km

– Typical Loaded truck weight: 15,455 sqm/truck + 15.6000 squm $\times$ 1.5 $\times$ 0.94/truck = 37,451 kg/truck
– 0.641 truck/ha

(8) E = (1000 km $\times$ 37,451 kg/truck $\times$ tonne/1000 kg $\times$ 0.641
      truck/ha) $\times$ 0.204 kg CO_2 eq/tonne-km
(9) E = 4,897.24 kg CO_2 eq/ha (Table 1)

Table 1. Summary of Geomembrane HDPE 1.5 mm carbon footprint

Process step	kg CO_2 eq/Ha	Assumptions
Manufacturing geomembrane HDPE 1.5 mm	25,944	From ICE 1.6a (polyethylene) = 1.6 tonnes CO_2/tonne PE
Transport to job site	4,897	1,000 km form Manufacturing Plant to Job Site
Total	30,841	kg CO_2 eq/ 10,000 m^2

1.2 Features of Compacted Clay Liners and Its Carbon Footprint

Compacted clay liners have been historically used as barrier layers in water lagoons and waste containment facilities. Common regulatory requirements for compacted clay liners are a minimum thickness of 0.6 m, with a maximum hydraulic conductivity of 1×10^{-11} m/s.

1.2.1 The Process
Clay at the borrow source is excavated using standard construction equipment, which also loads the material onto tri-axle dump trucks for transport to the job site. Each truck is assumed to have a capacity of 15 m^3 of loose soil. Using a compaction factor of 1.38, it is estimated that over 550 truckloads of soil would be needed to construct a 0.6-m thick compacted clay liner over a 1-ha area.

The distance from the borrow source to the job site is, of course, site-specific and can vary greatly. For the purposes of this analysis, a distance of 16 km (10 miles) was assumed. Since transport from the clay borrow source and the job site is such a large component of the overall carbon emissions, the sensitivity of the overall carbon footprint to changes in this site-specific variable is investigated later in this study (Table 2).

Table 2. Summary of compacted clay liner carbon footprint

Process step	kg CO_2 eq/Ha	Assumptions
Excavate soil at borrow source	2,656	CAT 329 Excavator, Operation 40 h/ha. Assume 24.5 l/h fuel consumption, based on medium work application an medium engine load factor (CAT performance handbook)
Haul clay to job site	93,527	**Assume site is 16 km from borrow source**, and 552 truckloads (each carrying 15 m^3 of clay) are needed to cover 1 ha
Construct clay liner	2,786	Operation 40 h/ha. Assume 25.7 l/h diesel fuel consumption
CAT D6 bulldozer	4,553	Operation 40 h/ha. Assume 42 l/h diesel fuel consumption
CAT 815 sheepsfoot compactor	4,553	Operation 40 h/ha. Assume 42 l/h diesel fuel consumption
CAT 815 smooth drum compactor	1,518	Operation 40 h/ha. Assume 14 l/h diesel fuel consumption
38,000 l water truck		
Total	109,593	Kg CO_2 eq/10,000 m^2 lined area with compacted clay

2   Conclusions

Considering all above mentioned, once again it is being showed the huge advantage to use geosynthetics instead traditional materials. When it comes down to waterproofing the Geomembrane ALVATECH HDPE 1.5 mm is by far the best choice because not only ensure the outstanding features in the long term (high chemical resistance and strong mechanical properties) but also is much more environmentally friendly solution.

It has 3 times less carbon footprint, not even have good quality clay at 16 km from borrow source to job site, comparing with geomembrane HDPE supplied by truck at 1.000 km from manufacturing plant.

In the end that is why, the geomembrane ALVATECH HDPE is being installed in wide range application for covering and ground protection in high demanding industries as mining, oil, landfills, waste water treatment and irrigation lagoons.

References

Franklin Associates: Final Report: LCI Summary for PLA and PET 12-Ounce Water Bottles (2007). http://www.petresin.org/pdf/FranklinPETPLAlifecycleanal_12-oz.pdf. Accessed Oct 2018

US EPA: U.S. Environmental Protection Agency. Mandatory reporting of greenhouse gases, final rule. Washington, DC (2009)

Uniaxial GEOGRID Pullout Mechanism Characterization from Calcareous Sand Backfill

Ahmad M. Abdelmawla[1]([⊠]), Yasser A. Hegazy[2],
Ashraf A. Al-ashaal[1], Sherif A. Akl[2], and Hesham K. Ameen[2]

[1] National Water Research Center, Cairo, Egypt
A.M.Abdelmawla@gmail.com
[2] Faculty of Engineering, Cairo University, Giza, Egypt

Abstract. There are two primary failure mechanisms associated with Reinforced Soil Systems (RSS) walls using geogrids including rupture and pullout. The failure mechanism with relevance to this research is the pullout failure. This type of failure mechanism depends on the interaction between the backfill soil and reinforcement. In order to select a type of backfill soil or reinforcement that will provide sufficient interaction strength, the soil reinforcement interaction should be tested and characterized using pullout tests. Siliceous sands have been widely used as a backfill soil in MSE walls. Also, most published researches addresses the geogrid pullout resistance embedded in siliceous sands. geogrid pullout design empirical equations and parameters in design codes such as AASHTO are based on using siliceous sand as a backfill material. In this paper, the aim is to evaluate the interaction between calcareous sand as a backfill soil and two types of geogrids including High Density Poly Ethylene (HDPE) and polyester, performing and analyzing laboratory pullout test results. Pullout tests were performed at vertical stresses of 100, 150 and 200 kPa corresponding to approximate wall heights of 5 to 11 m. The results of pullout test are used to define geogrid pullout resistance parameters from calcareous sand. The results of this study show that: (1) Pullout resistance force for HDPE geogrid embedded in siliceous sand are higher than those embedded in calcareous sand by 17%, (2) Pullout resistance force for polyester geogrid embedded in calcareous sand are higher than those embedded in siliceous sand by 6%, (3) The pullout resistance factor for polyester geogrids embedded in calcareous sand is greater by 4% than the same factor for geogrids embedded siliceous sand. However the pullout resistance factor for HDPE geogrid embedded in calcareous sand is smaller by 16% than the same factor for geogrids embedded in siliceous sand.

1 Introduction

Some laboratory of tests can be performed to define the interaction parameters of soil with geosynthetics as shown in Fig. 1, including biaxial, pullout direct shear and modified direct shear tests, pullout test is the one used in for this research. The experimental work for this research took place in the soil mechanics and foundations

© Springer Nature Switzerland AG 2019
M. Meguid et al. (Eds.): GeoMEast 2018, SUCI, pp. 248–257, 2019.
https://doi.org/10.1007/978-3-030-01944-0_20

laboratory at construction research institute (CRI) of national water research center (NWRC) in Egypt.

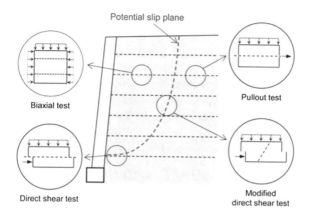

Fig. 1. Testing types of reinforced soil

In this paper, the focus is to evaluate the interaction between calcareous sand as a backfill soil and two types of geogrids including HDPE and polyester using large-scale pullout tests.

Calcareous sands typically contain high carbonate contents resulting in high crushability. Calcareous soils are different from siliceous soils in their formation and structure of their particles. Figure 2 shows the structure of Egypt north beach calcareous sands that were used to perform laboratory tests of this paper.

Fig. 2. Egypt north beach calcareous sand. (Adopted from Salehzadeh et al. 2008)

The pullout resistance of the soil reinforcement is defined by the ultimate tensile load required to generate sliding outward of the reinforced soil zone. The pullout resistance of the reinforcement is mobilized through one of or the two basic soil-reinforcement interaction mechanisms including interface friction, and passive soil resistance against transverse elements of reinforcements such as bar mats, wire meshes, or geogrids. The soil-to-reinforcement relative movement mobilizing the design tensile force depends on the load transfer mechanism, the extensibility of the reinforcement material, the soil type, and the confining pressure. The effective pullout resistance is estimated using Eq. (1):

$$P_r = F^* \cdot L_e \cdot \alpha \cdot \sigma_v \cdot R_c \qquad (1)$$

Where, P_r = the pullout resistance, F^* = pullout friction factor, L_e = length of reinforcement in resisting zone, α = scale effect correction factor, σ_v = unfactored vertical stress at the reinforcement level in the resistant zone and R_c = reinforcement coverage ratio.

2 Laboratory Pullout Testing Program

The geogrid pullout resistances using siliceous and calcareous sands are compared using large scale pullout laboratory tests compacted at relative density of 88–90% using applied vertical stresses of 100, 150 and 200 kPa corresponding to approximate wall heights of 5 to 11 m.
The tests were performed in a large pullout apparatus, shown in Fig. 3 including:

- driving system (electric motor with screw bars);
- pneumatic air pressure (pressurized air bag – air compressor);
- Steel box (2.00 m length × 1.00 m width × 1.00 m height).

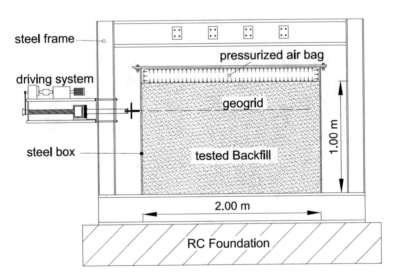

Fig. 3. Pullout large scale testing box

2.1   Properties of Used Calcareous Sands

Calcareous sands used in this study are brought from Egypt north coast and are medium to fine poorly graded sands as shown in Fig. 4 and Table 1. The sand is collected 100 m inland away from the shoreline where shells and gravels are avoided.

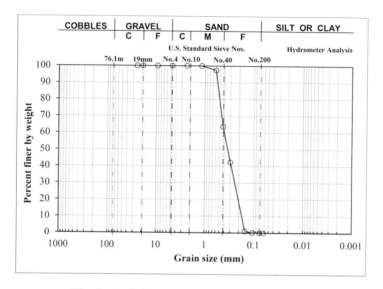

Fig. 4. Gradation curve of used calcareous sands

Table 1. Selected Particle sizes and classification coefficients of used calcareous sands

Particle size/Coefficient	Calcareous sand
D10 (mm)	0.174
D30 (mm)	0.244
D60 (mm)	0.401
CU (coeff. of uniformity)	2.297
Cc (coeff. of curvature)	0.852

The minimum and maximum density test results show that calcareous sand has a small range of change in voids ratio. The maximum and minimum densities of used calcareous sands are between 1.75 and 1.63 t/m^3 corresponding to minimum and maximum void ratios of 0.5 and 0.6 approximately (Table 2).

The pullout tests were conducted using calcareous sand compacted in dry condition with a water content of 0.33%.

A direct shear box test was used to calculate the calcareous sand internal angle of friction and the results are shown in Fig. 5. Shear box test curve, indicating that the

Table 2. Maximum and minimum dry densities test results

Property	Calcareous sand
Minimum density (γ_{min} t/m^3)	1.63
Maximum voids Ratio (e_{max} %)	59.91
Maximum density (γ_{max} t/m^3)	1.75
Minimum voids ratio (e_{min} %)	49.30

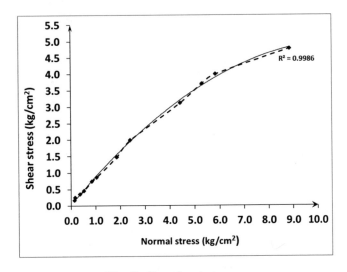

Fig. 5. Shear box test curve

higher the normal stress the more crushing occurs to the sand particles. Which resulted in reduction of the angle of friction.

Equation (2) is used to estimate the secant frictional angle, $\phi = \tan^{-1}(\tau/\sigma)$ at different normal stresses during pullout tests from calcareous sand at different normal stresses less than 8.5 kg/cm^2 (850 kPa).

$$\tau = -0.0402\sigma^2 + 0.8968\sigma + 0.0301 \ (\text{kg/cm}^2) \tag{2}$$

where, τ = shear resistance and σ = normal stress

2.2   Properties of Used Geogrids

Two types of geogrids were used in this study including HDPE uniaxial geogrid, (RE 510) and Stratagrid uniaxial geogrid (SG 350) as summarized in Table 3 for a length of tested specimens of 1.75 m.

Table 3. Strength data of used geogrids

GEOGRID type	HDPE	Polyester
Tensile strength (kN/m width)	44	73
Strain at maximum tensile strength	11.0 ± 3.0%	16%

3   Results and Analysis

3.1   Results

All pullout tests were performed until rupture of geogrid or obtaining a total horizontal displacement of 75 mm whichever comes first according to ASTM D6706-01 to assure confinement of geogrid specimens in the backfill soil. The maximum pullout forces (pullout resistances) obtained in the tests are reported in Table 3.

Figures 6 and 7 show measured pullout forces versus the displacements of the geogrid at the clamp. Figures 6 and 7 show results of tests performed on HDPE GEOGRID and polyester geogrid specimens respectively using different confining pressures and backfill soils.

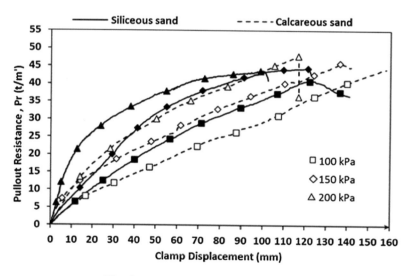

Fig. 6. Pullout curves for HDPE geogrid

The trend of pullout resistance with displacement is similar for siliceous and calcareous sands back fills at normal pressure of 100 kPa as shown on Fig. 6. Pullout curves for HDPE geogrid. The pullout resistance exhibits a linear behavior for both backfill types until reaching rupture of geogrid at nearly 40–45 kN/m and displacements in range of 140–160 mm (Table 4).

At normal pressure of 150 kPa, using siliceous sand backfill, rupture of geogrid was reached at displacement of 124 mm and pullout force of 44 t/m.

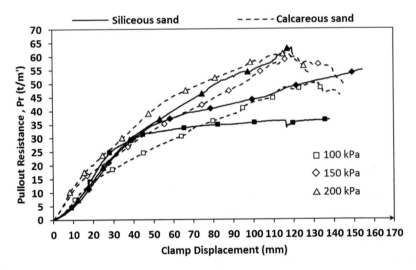

Fig. 7. Pullout curves for Polyester geogrid

Table 4. Pullout test results, Pr (t/m)

Backfill soil	GEOGRID type	Pullout force (t/m)		
		Surcharge pressure 100 kPa	Surcharge pressure 150 kPa	Surcharge pressure 200 kPa
Siliceous sand	HDPE	29.79	38.58	41.70
	Polyester	34.44	40.12	46.25
Calcareous sand	HDPE	23.90	31.71	37.33
	Polyester	34.79	42.22	50.06

At 200 kPa normal pressure, the rupture of geogrid was reached at 100 mm and 120 mm for siliceous and calcareous sands backfills.

For polyester geogrid at surcharge pressure of 100 kPa, siliceous and calcareous sand backfills have very different behaviors during pullout test as shown on Fig. 7. Siliceous sand has an increasing rate of resistance until reaching a plateau at 30 kN/m, which is defined as a pullout failure. But with the calcareous sand and due to the higher contact between geogrid and backfill the resistance continue to increase until reaching partial rupture of the geogrid then the resistance decreases rapidly with no defined plateau as in siliceous sand.

At surcharge pressure 150 kPa and according to as shown on Fig. 7 both siliceous and calcareous sand have the same trend of resistance against pullout until displacement of nearly 50 mm. The rate of increase for siliceous sand backfill becomes smaller than the rate at the beginning. For calcareous sand backfill the rate of increase, still almost the same from the beginning of the test until reaching the maximum pullout resistance at displacement of 120 mm and partial rupture of the geogrid then the rate of resistance decreases until the test end.

Pullout testing at surcharge pressure 200 kPa and as shown on Fig. 7 reveals that both siliceous and calcareous sands have the same trend of resistance against pullout until reaching partial rupture of geogrid at displacement of 115 mm and pullout force of 60 kN/m. The difference appears after the partial rupture. For the siliceous sand, the resistance continues to increase after partial rupture of geogrid while for the calcareous sand the resistance decreases after the partial rupture. The resistance increases for siliceous sand after the geogrid rupture due to dilation of sand particles that leads to increase of the friction resistance between sand and geogrid (soil hardening). While for calcareous sand, the contraction between sand particles due to a relatively high inter-particles void ratio leads to decrease in sand volume during test (soil softening); on contrary of the siliceous sand, the above leads to higher strains in calcareous sand until reaching same strength as siliceous sand.

3.2   Analysis

Figure 8 shows the trends of measured pullout forces with applied normal stresses.

Figure 8a indicate that embedding HDPE geogrids in a siliceous sand backfill resulted in greater pullout resistance than that using calcareous sand backfill.

However, Fig. 8b indicates that embedding polyester geogrids in a calcareous sand backfill resulted in a greater pullout resistance than using a siliceous sand backfill.

Figure 9 shows that correlation between the back calculated pullout resistance factor (F*) of siliceous and calcareous sands backfills in case of using HDPE and polyester geogrids.

A linear correlation was established between F*(calcareous sand) and F*(siliceous sand) using curve fitting as shown in Eq. (3), where m_{cal} is a factor correlates the F* factors, and its values are summarized in Table 5.

$$m_{cal.} = \frac{F * (calcareous \, sand)}{F * (Siliceous \, Sand)} \tag{3}$$

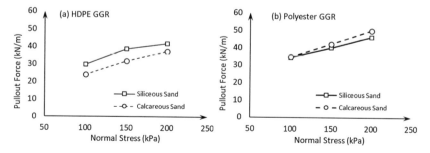

Fig. 8. Force versus applied vertical stresses

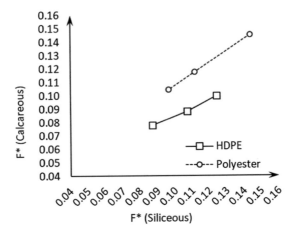

Fig. 9. Relation between Pullout resistance factor (F*) for siliceous and calcareous sand

Table 5. Calculated values for $m_{cal.}$ factor

GEOGRID type	m_{cal}
HDPE	0.829
Polyester	1.038

According to previous analysis, the pullout resistance factor (F*) for calcareous sand with HDPE decreased by nearly 17% than that for siliceous sand. This may be due to the smooth nature of HDPE geogrid surface and despite of the high internal friction angle of calcareous sand, but the high friction angles of carbonate sands are not the result of a dilatant behavior, and the soil is softer and more compressible after grain crushing and volumetric change induced by normal or shearing stresses. As softening behavior of calcareous sand during shearing process leads to large strains before reaching maximum resistance as the siliceous sand. So this is why at the same strain level predefined by ASTM standard, the siliceous sand has a higher pullout resistance.

While for polyester geogrid, due to its rough surface nature that mobilizes higher friction resistance with crushed calcareous sand sooner than that of HDPE geogrid. So, due to the higher internal friction angle of the interface, the pullout friction factor (F*) for calcareous sand with polyester increases by almost 4% than that of siliceous sand with the same geogrid.

4   Conclusions

The main conclusions of this study are summarized below:

- The pullout resistance results, as well as the pullout resistance coefficient, was found to be considerably higher for the polyester geogrid compared to the HDPE. This was found to be the case for both siliceous sand and calcareous sand.

- Pullout resistance force for HDPE geogrid embedded in siliceous sand are greater than those embedded in calcareous sand by 17%. Due to the softening and crushing effect of calcareous sand during pullout of geogrid. Also since the smooth surface of HDPE geogrid, it does not perform high interlocking with calcareous sand and have more interaction with siliceous sand.
- Pullout resistance force for polyester geogrid embedded in calcareous sand are greater than those embedded in siliceous sand by 6%. Due to the unique nature of polyster geogrid (rough surface) and due to the angularity of crushed calcareous sand grains.
- The pullout resistance factor for polyester geogrids embedded in calcareous sand is greater by 4% than the same factor for geogrids embedded siliceous sand.
- The pullout resistance factor for HDPE geogrid embedded in calcareous sand is smaller by 16% than the same factor for geogrids embedded in siliceous sand.

Reference

Salehzadeh, H., Hassanlourad, M., Procter, D.C., Merrifield, C.M.: Compression and extension monotonic loading of a carbonate sand. Int. J. Civil Eng. **6**, 266–274 (2008)

Author Index

© Springer Nature Switzerland AG 2019
M. Meguid et al. (Eds.): GeoMEast 2018, SUCI, pp. 259–260, 2019.
https://doi.org/10.1007/978-3-030-01944-0

Printed in the United States
By Bookmasters